Y0-BWU-112

Walch Hands-on Science Series

Arthropods

by Susan Appel

illustrated by Lloyd Birmingham

Project Editors:
Joel Beller
Carl Raab

User's Guide
to
Walch Reproducible Books

As part of our general effort to provide educational materials that are as practical and economical as possible, we have designated this publication a "reproducible book." The designation means that the purchase of the book includes purchase of the right to limited reproduction of all pages on which this symbol appears:

Here is the basic Walch policy: We grant to individual purchasers of this book the right to make sufficient copies of reproducible pages for use by all students of a single teacher. This permission is limited to a single teacher and does not apply to entire schools or school systems, so institutions purchasing the book should pass the permission on to a single teacher. Copying of the book or its parts for resale is prohibited.

Any questions regarding this policy or request to purchase further reproduction rights should be addressed to:

Permissions Editor
J. Weston Walch, Publisher
P.O. Box 658
Portland, Maine 04104-0658

Acknowledgments

My special thanks to the following people, without whom this book would never have been written: Carl Raab for his encouragement and support; Joel Beller, a master teacher and editor, for offering wonderful suggestions; Judy Goldberg for her way with words and critical eye; Lloyd Appel for being a sounding board and tolerating our basement arthropod laboratory; and my students, whose exciting ideas for projects have stimulated me to learn more about arthropods.

To my grandson, Matthew Alex, who made his first sounds during the time this book was written. Welcome to the chorus.

1 2 3 4 5 6 7 8 9 10
ISBN 0-8251-3761-6
Copyright © 1998
J. Weston Walch, Publisher
P. O. Box 658 • Portland, Maine 04104-0658
Printed in the United States of America

Contents

To the Teacher

This book on arthropods is one of a series of books devoted to hands-on science activities for middle school and early high school students. The activities in this book are intended as a supplement to the textbook and curriculum. Their purpose is to provide enrichment and to engage students' curiosity about living organisms. They may also be used as a springboard for science projects or club activities and to involve other members of the family in science as a process.

The thrust of these studies is to observe arthropods in natural or laboratory environments and to teach youngsters to respect and appreciate biodiversity. This book is not about dissection or experiments involving animal sacrifice but rather about assuming the role of behavioral scientists and treating animals humanely.

The organisms that have been selected for these activities have several traits in common. They are harmless to humans and easy to obtain and maintain in a school laboratory. Although these organisms are readily available from biological supply houses, some, such as pill bugs and sow bugs, may be collected locally. To help minimize expense and storage space, organisms such as crickets, and fruit flies may be used for more than one activity.

Most of the equipment used here is simple and inexpensive. Small plastic tanks covered with screens will provide housing for most of the organisms, and shoe boxes may be used as arenas for various experiments. It will be helpful, however, if one or more stereomicroscopes are available for making more detailed observations of these animals.

The activities range in difficulty as well as time needed to complete them. Some can be completed in one class period (Terrestrial Crustaceans: Pill Bugs and Sow Bugs); others require several weeks (Incomplete Metamorphosis: A Cricket's Life Cycle). Although most can be done in class, some must be completed out-of-doors (Finding Arthropods: The Backyard Laboratory). Choose the activities that meet the needs of your students.

Each activity has suggestions for follow-up exercises as well as a teacher resource page. The latter contains helpful hints, adaptations for high and low achievers, Internet tie-ins, and quiz questions. Instructional objectives and National Science Standards are included to help you meet state and local expectations. Above all, these activities are intended to open students' eyes to the wonders of arthropods. Enjoy!

Finding Arthropods: The Backyard Laboratory

TEACHER RESOURCE PAGE

INSTRUCTIONAL OBJECTIVES

Students will be able to

- record observations of living organisms.
- draw and label organisms.
- classify organisms based on their anatomy.
- record data in a data table.
- test hypotheses.
- draw conclusions based on data.

NATIONAL SCIENCE STANDARDS ADDRESSED

Students demonstrate understanding of

- structure and function of living systems.
- diversity and adaptation of organisms.
- big ideas and unifying concepts, such as order and organization.

Students demonstrate scientific inquiry and problem-solving skills by

- identifying variables.
- using evidence from reliable sources to develop explanations and models.
- working in teams to collect and share information and ideas.

Students demonstrate competence with the tools and technology of science by

- using traditional laboratory equipment to observe and measure organisms.
- recording data in a variety of formats.
- acquiring information from multiple sources, such as print, the Internet, and experimentation.

Students demonstrate effective scientific communication by

- representing data in multiple ways, such as numbers, tables, and drawings.
- explaining a scientific concept or procedure to others.

MATERIALS

Each group of four students will need

- Hand lens
- Large spoon or digging tool
- Jar with ring of petroleum jelly around inside rim and holes in cover to collect arthropods
- Petri dish
- Stereomicroscope
- Berlese-Tullgren apparatus (see Figure 1)
- Plastic gloves
- Metric ruler

HELPFUL HINTS AND DISCUSSION

Time frame: Two periods of instruction
Structure: Groups of four students
Location: First period outdoors, second period in class

In this activity, students will explore their local surroundings to observe and identify the classes of organisms belonging to the phylum *Arthropoda.* This is the most successful and most diverse group of animals on our planet and contains approximately one million different species. All arthropods have a hard, nonliving exoskeleton made of chitin, jointed legs, and segmented body parts. They shed their skeletons (molt) when they grow and eventually will produce a new skeleton.

When introducing this activity, emphasize to the students that they will be observing living organisms in their natural habitat and should treat these organisms humanely. Before beginning the outdoor observation, discuss the classes of arthropods found in Table 1 of the student activity page. You might want to present some living arthropods or photographs and ask your students to draw and classify them according to this table.

Stress the need for accurate field drawings, including number of body segments, antennae, and legs.

(continued)

Adaptations for High and Low Achievers

High Achievers: These students should help lower achievers by working with them in a group. Also, encourage these students to perform additional experiments based on follow-up activity 2, and to identify the genus and species of arthropods based on follow-up activity 3.

Low Achievers: Provide a glossary and/or reference material for the italicized terms in this activity. Heterogeneous groupings will enable students of higher ability to assist those having difficulty carrying out the activity. Encourage the good artists in this group to produce artwork based on their observations.

Scoring Rubric

Full credit should be given to students who use complete sentences to correctly answer the questions and accurately complete the two data tables provided. Extra credit should be given to students who complete any of the follow-up activities.

Internet Tie-ins

www.birminghamzoo.com/ao/arthrop.htm (Animal Omnibus)
http://azstarnet.com/~sasi/ (Sonoran Arthropod Studies Institute)

Quiz

Answer each of the following questions in one or more complete sentences.
You find an unknown animal outside the school. You think that it might be an arthropod.

1. What would you need to know about the animal before deciding that it is an arthropod?
2. Describe two ways you could tell whether the animal belongs to the phylum *Insecta* or the phylum *Arachnida.*

Name____ Date ____________________

Finding Arthropods: The Backyard Laboratory

STUDENT ACTIVITY PAGE

Welcome to the world of arthropods, the most successful group of animals on our planet. Two out of every three organisms belong to this *phylum.* There are approximately one million different *species* of arthropods; most are insects. All arthropods all have a hard, nonliving *exoskeleton,* jointed legs, and segmented body parts. Because they shed their exoskeletons when they grow, arthropods tend to be small. Some are microscopic.

Arthropods are everywhere—on land and in the sea and air. Today you will look for them in your own backyard or neighborhood. You will learn something about their lifestyles and decide which *class* they are in.

Remember that the arthropods you find are alive and must be treated humanely. Take time to observe them in their natural habitat. Do not make any sudden noises or movements. If you must move the animals, be careful not to harm them and give them time to recover. Although most arthropods that you find will not harm humans, make sure you handle them with care. Because some people may be allergic to certain arthropods, it is advisable to wear gloves if you touch any of these animals. At the end of this activity, return the organisms to their habitat and make sure that you wash your hands thoroughly.

- Hand lens
- Large spoon or digging tool
- Jar with ring of petroleum jelly around inside rim and holes in cover to collect arthropods
- Petri dish
- Stereomicroscope
- Berlese-Tullgren apparatus (see Figure 1)
- Plastic gloves
- Metric ruler

You will be working in groups of four. This will enable you to share your observations. You will learn more if you encourage input from all the members of the group. Work as a team to answer the questions and to record your data and answers in the Data Collection and Analysis section.

1. Find an outdoor habitat where arthropods can be found. This may be a school yard, garden, backyard, forest, vacant lot, or crack in a sidewalk. In short, almost any land environment is appropriate. Depending on the location you select, you may not be able to complete all of the following activities. Do as many as possible.
2. For each location in steps 3 to 6, do the following:
   (a) identify the class of each arthropod you find according to the characteristics in Table 1.
   (b) Describe the motion of each arthropod.
   (c) Describe any other activities you observe.
   (d) If possible, measure the arthropod to the nearest 0.1 cm.
   (e) Draw the arthropod and label as many structures as possible.

(continued)

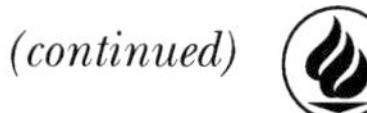

Name____ Date ______________________

Finding Arthropods: The Backyard Laboratory *(continued)*

STUDENT ACTIVITY PAGE

Table 1: Classification of Arthropods

Class	Characteristics	Examples
Crustacea	2 pairs of antennae	lobsters, crabs, water fleas, sow bugs
Chilopoda	1 pair of antennae many body segments 1 pair of legs per segment	centipedes
Diplopoda	similar to Chilopoda, but 2 pairs of legs per segment	millipedes
Arachnida	no antennae 2 body regions 4 pairs of legs	spiders
Insecta	1 pair of antennae 3 body regions 3 pairs of legs	flies, ants, bees, cockroaches

3. Look under a rock, first with the naked eye and then with a hand lens. Describe what you see.
4. Using your digging tool, collect approximately half a jar of soil from under the rock. Bring it back to class for further examination.
5. Look on the leaves, trunk, and branches of a tree and on the surface of the soil around the tree. How do the arthropods in these locations compare with those under the rock?
6. Look for arthropods in other locations, such as a crack in a sidewalk, on or under fallen leaves, or near any food that might have dropped in the area. How do these arthropods compare with others you found?
7. When you get back to the classroom, place some of your soil in a petri dish, and observe it with a stereomicroscope. What new organisms do you see?
8. Place the rest of the soil in the wire basket in a Berlese-Tullgren apparatus (Figure 1). Turn on the lamp. Describe and identify any soil arthropods that drop through the bottom of the funnel into the beaker below.
9. With your group, decide how to present your results to the class.
10. When you have finished, clean up your lab station and return your organisms to their natural habitat.

Figure 1: Berlese-Tullgren Apparatus

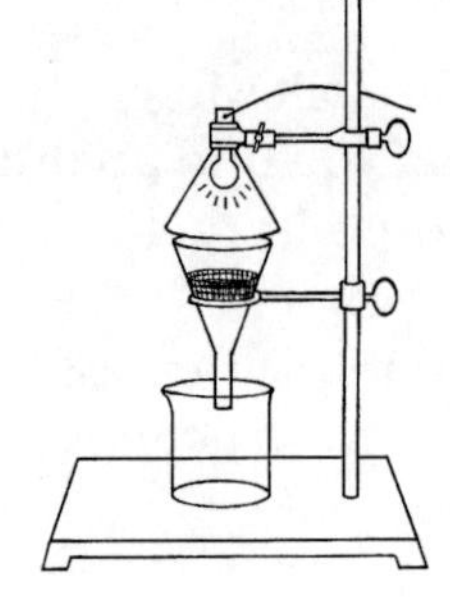

(continued)

Name________________________________ Date ____________________

Finding Arthropods: The Backyard Laboratory *(continued)*

STUDENT ACTIVITY PAGE

DATA COLLECTION AND ANALYSIS

1. Complete the following table for each organism you found.

Class and Common Name	Drawing and Size (cm)	Location	Description of Activities

2. Complete the following summary table.

Classes of Arthropods

Location	# of Crustaceae	# of Chilopoda	# of Diplopoda	# of Arachnida	# of Insecta

3. On the back of this sheet, draw a picture of the organism(s) that you observed under the microscope. Label as many structures as possible. Determine the class of the organism and if possible state its common name.
4. Explain how the arthropods you collected in the beaker at the bottom of the Berlese-Tullgren apparatus were similar to the arthropods at the surface of the soil. How were they different?

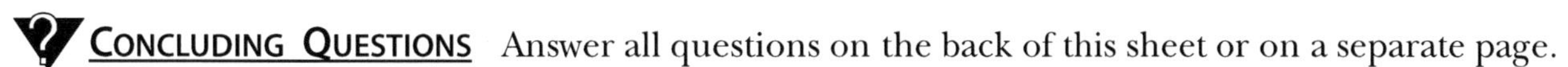

CONCLUDING QUESTIONS

Answer all questions on the back of this sheet or on a separate page.

1. Explain why you might have found different types of arthropods at different times of the day, or different seasons of the year.
2. How would you determine whether an arthropod that you found in the soil was an insect or a crustacean?

FOLLOW-UP ACTIVITIES

1. Refer to your notes and drawings of the insects you found. How did the locomotion of insects that were on or aboveground compare with the locomotion of insects that were under a rock?
2. Design a controlled experiment to test the following hypothesis: The light at the top of the Berlese-Tullgren apparatus causes the arthropods to burrow deeper into the soil and eventually fall through the screen. Try your experiment.
3. Using your textbook, other printed materials, or the Internet, identify the scientific and/or common names of as many of your arthropods as possible.

© 1998 J. Weston Walch, Publisher

Finding Arthropods: The Aquatic Laboratory

TEACHER RESOURCE PAGE

Instructional Objectives

Students will be able to:

- record observations of living organisms.
- draw and label organisms.
- classify organisms.
- record data in a data table.
- test hypotheses.
- draw conclusions based on data.

National Science Standards Addressed

Students demonstrate understanding of

- structure and function of living systems.
- diversity and adaptation of organisms.
- big ideas and unifying concepts, such as order and organization.

Students demonstrate scientific inquiry and problem-solving skills by

- identifying variables.
- using evidence from reliable sources to develop explanations and models.
- working in teams to collect and share information and ideas.

Students demonstrate competence with the tools and technology of science by

- using traditional laboratory equipment to observe and measure organisms.
- recording data in a variety of formats.
- acquiring information from multiple sources, such as print, the Internet, and experimentation.

Student demonstrate effective scientific communication by

- representing data in multiple ways, such as numbers, tables, and drawings.
- explaining a scientific concept or procedure to others.

Materials

Each group of four students will need

- Hand lens
- Large spoon or digging tool
- Net or net bag
- Plastic bucket with handle
- Metric ruler
- Stereo- and/or compound microscope
- Slides
- Coverslips
- Medicine dropper
- Small plastic spoon or scoop
- Culture dish

Helpful Hints and Discussion

Time frame: Two periods of instruction
Structure: Groups of four students
Location: First period outdoors, second period in class

In this activity, students will explore local aquatic communities to observe and identify the classes of organisms belonging to the phylum *Arthropoda.* In addition to marine organisms found in oceans, aquatic organisms are also found in freshwater communities and wetlands. Depending on your location, you may choose to do some rather than all of the activities contained here.

Before beginning the outdoor observation, discuss the classes of arthropods found in Table 1. Distribute copies of the table. You might want to present some living arthropods or photographs and ask your students to draw and classify them according to this table. Discuss the importance of accurate field drawings, including the number of body segments, antennae, and legs.

When introducing this activity, emphasize to the students that they will be observing living organisms in their natural habitat and should treat the organisms humanely. If your students are planning to maintain an aquarium, as suggested in follow-up activity 2, make sure that they have done the necessary research.

(continued)

Adaptations for High and Low Achievers

High Achievers: These students should help lower achievers by working with them in a group. Encourage these students to identify the genus and species of arthropods based on concluding question 2.

Low Achievers: Provide a glossary and/or reference material for the italicized terms in this activity. Heterogeneous groupings will enable students of higher ability to assist those having difficulty carrying out the activity. Encourage the artists in this group to produce artwork based on their observations.

Scoring Rubric

Full credit should be given to students who use complete sentences to correctly answer all questions and accurately complete the two data tables provided. Extra credit should be given to students who complete any of the follow-up activities.

www.birminghamzoo.com/ao/arthrop.htm (Animal Omnibus)
http://azstarnet.com/~sasi (Sonoran Arthropod Studies Institute)

Answer each of the following questions in one or more complete sentences.

You find 10 shells on the beach. After close examination, you realize that only one is from an arthropod.

1. How could you tell that the shell is from an arthropod?
2. Describe two ways you could tell whether the shell belonged to an organism from the class *Crustacea* or the class *Insecta.*

Name____ Date ____________________

Finding Arthropods: The Aquatic Laboratory

Student Activity Page

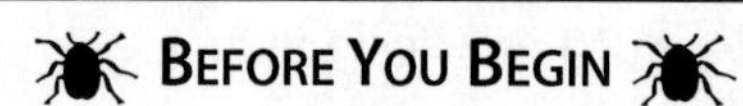

Although you are probably most familiar with the arthropods that live on land, many are found in aquatic communities. Many areas of the world contain aquatic communities because nearly 75 percent of the globe is covered by oceans. The organisms that live in the salty oceans are *marine* organisms. In addition to the marine communities, there are many *freshwater* communities found in *rivers, streams, ponds,* and *lakes.*

Some places are flooded for all or part of the year. These areas are called *wetlands.* Examples of wetlands are *marshes, bogs,* and *swamps.* These wetlands contain many types of plants and animals.

In this activity, you will examine arthropods found in your local aquatic communities. Depending on where you live, you may be investigating any body of water, ranging from a puddle to an ocean. Remember that the arthropods that you find are alive and must be treated humanely. Also, make sure that you take proper precautions for your own safety. To avoid slipping, wear shoes with rubber soles. Do not make any sudden noises or movements. If you must move the animals, be careful not to harm them and give them time to recover. Although most arthropods that you find will not harm humans, make sure you handle them with care. At the end of the activity, return the organisms to their habitat and make sure that you wash your hands thoroughly.

Materials

- Hand lens
- Large spoon or digging tool
- Net or net bag
- Plastic bucket with handle
- Metric ruler
- Stereo- and/or compound microscope
- Slides
- Coverslips
- Medicine dropper
- Small plastic spoon or scoop
- Culture dish

Procedure

1. With the help of your teacher, find an aquatic habitat containing arthropods. This can range from a puddle to the ocean. Depending on the location, you may not be able to complete all of the following activities. Do as many as possible.

2. For each location in steps 3 to 6, do the following:
   (a) Look for animals that might be arthropods.
   (b) Identify the class of each arthropod you find according to the characteristics in Table 1.
   (c) Describe the motion of each arthropod.
   (d) Describe any other activities you observe.
   (e) If possible, measure the arthropod to the nearest 0.1 cm.
   (f) Draw the arthropod and label as many structures as possible.
   (g) Collect some water and sand or soil from the shallow region of each location. Bring this back to the laboratory for further examination.

(continued)

© 1998 J. Weston Walch, Publisher

Finding Arthropods: The Aquatic Laboratory *(continued)*

3. *At the Seashore*
   (a) Walk along the shoreline and look for arthropods. If you see holes in the sand as the tide goes out, there may be an arthropod in the sand. Using your digging tool, try to uncover it.
   (b) At low tide, look among the rocks and in tidepools. *Be careful not to get trapped in these areas at high tide.* How do these arthropods compare with the ones at the shoreline?
   (c) Dip your net into shallow water at various locations and try to identify the arthropods that you capture. After you have examined the organisms, return them to their habitat.
4. *At a brook or stream*
   (a) Examine the surface of the running water. You may find arthropods that seem to be walking or striding on the surface. Closely examine their structure to determine why they don't sink.
   (b) Look under a rock. How do the arthropods that you find here compare with those that were on the surface of the water in terms of size and body structure, especially in the limbs, body segments, and antennae?
   (c) Dip your net into the water at various locations and try to identify any arthropods you find. When you have finished your examination, return the organisms to their habitat.
5. *At a lake, pond, or puddle*
   (a) Look for arthropods on the surface of the water. Why don't they sink?
   (b) At the shoreline, look for both aquatic and terrestrial arthropods. Look under a rock in very shallow water and then under a rock on land. How do the organisms under the two rocks compare with each other in terms of size and body structure, especially the limbs, body segments, and antennae?
   (c) Dip your net into shallow water and identify any arthropods that you find. When you have completed your examination, return the organisms to their habitat.
6. *At wetlands (marshes, bogs, or swamps)*
   (a) You are likely to find arthropods among the *vegetation* in wetlands. Look for arthropods around the roots and on the leaves of plants as well as on the surface of the water. How do the arthropods compare with each other?
   (b) Dip your net into the water and identify any arthropods you find. When you have finished your examination, return the organisms to their habitat.
7. When you get back to the laboratory, place some of your water in a culture dish and examine it under a stereomicroscope. What organisms do you see?
8. Using some of the water in the culture dish, make a slide and examine it under a compound microscope. What new organisms do you see?
9. With your group, decide how to present your results to the class.
10. When you have finished, clean up your lab station. Unless you are planning to complete follow-up activity 2, return your organisms to their natural habitat.

(continued)

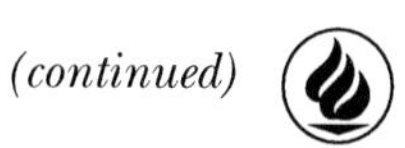

Name______________________ Date ______________

Finding Arthropods: The Aquatic Laboratory *(continued)*

STUDENT ACTIVITY PAGE

Table 1: Classification of Arthropods

Class	Characteristics	Examples
Crustacea	2 pairs of antennae	lobsters, crabs, water fleas, sow bugs
Chilopoda	1 pair of antennae many body segments 1 pair of legs per segment	centipedes
Diplopoda	similar to Chilopoda, but 2 pairs of legs per segment	millipedes
Arachnida	no antennae 2 body regions 4 pairs of legs	spiders
Insecta	1 pair of antennae 3 body regions 3 pairs of legs	flies, ants, bees, cockroaches

DATA COLLECTION AND ANALYSIS

1. Complete the following table for each organism you found.

Class and Common Name	Drawing and Size (cm)	Location	Description of Activities

2. Complete the following summary table.

Classes of Arthropods

Location	# of Crustacea	# of Chilopoda	# of Diplopoda	# of Arachnida	# of Insecta

(continued)

Name____ Date ____________________________

Finding Arthropods: The Aquatic Laboratory *(continued)*

STUDENT ACTIVITY PAGE

3. On the back of this page, draw a picture of the organism(s) that you observed under the microscope. Label as many structures as possible. Determine the class of the organism and if possible state its common name.

FOLLOW-UP ACTIVITIES

1. Refer to your notes and drawings of the arthropods you found. Select at least two organisms and describe how they are adapted to survive in their environment.
2. Set up an aquarium under your teacher's supervision. In it try to maintain the organisms that you brought back to class.

Under the Microscope: A View of Copepods

INSTRUCTIONAL OBJECTIVES

Students will be able to

- use a stereomicroscope to make observations of a living organism.
- record observations.
- record data in a data table.
- draw conclusions based on data.
- formulate hypotheses based on aspects of copepod behavior.
- draw and label a copepod.

NATIONAL SCIENCE STANDARDS ADDRESSED

Students demonstrate understanding of

- structure and function in living systems.
- regulation and behavior, such as response to environmental stimuli.
- diversity and adaptations of organisms.
- big ideas and unifying concepts, such as order and organization and cause and effect.

Students demonstrate scientific inquiry and problem-solving skills by

- identifying or controlling variables in experimental and nonexperimental research settings.
- using evidence from reliable sources to develop descriptions and explanations.
- working individually and in teams to collect and share information and ideas.

Students demonstrate competence with the tools and technologies of science by

- using traditional laboratory equipment to observe organisms and phenomena.

Students demonstrate effective scientific communication by

- representing data in multiple ways, such as technical writing and drawing.
- communicating in a form suited to the purpose and audience, such as by writing instructions that others can follow.
- recording and storing data.
- acquiring information from multiple sources such as print, the Internet, and experimentation.
- recognizing sources of bias in data, such as observer and sampling biases.

MATERIALS

Each pair of students will need

- Jar containing copepods in freshwater
- Plastic spoon or scoop
- Small culture dish
- Stereomicroscope
- Medicine dropper

HELPFUL HINTS AND DISCUSSION

Time frame: Single period of instruction
Structure: Pairs of two students
Location: In class

In this activity, students will use the naked eye and stereomicroscopes to observe copepods, which are tiny aquatic crustaceans. These organisms are inexpensive and may be ordered from biological supply houses. If available, an interesting copepod would be *Cyclops,* so named because of its single eye.

Freshwater copepods are more readily available and easier to maintain than marine copepods. Most are herbivores and will survive in a small aquarium containing aquatic plants. In their natural habitat, copepods are a major source of food for fish and other aquatic animals. In fact, they comprise most of the diet of some of the largest whales.

Before introducing the organism, make sure that your students know how to use the stereomicroscope. Show them how to determine the total magnification.

Copepods are very fast swimmers but may be captured easily in a small spoon. Suggest that your students collect only a half teaspoon of liquid with the organism and place this in the culture dish. The amount will not completely cover the bottom of the dish and will thus keep the copepod in the field of vision.

(continued)

Adaptations for High and Low Achievers

High Achievers: These students should help lower achievers by working with them in a group. Also, encourage these students to perform additional experiments based on follow-up activity 2.

Low Achievers: Provide a glossary and/or reference material for the italicized terms in this activity. Heterogeneous groupings will enable students of higher ability to assist those having difficulty carrying out the activity.

Scoring Rubric

Full credit should be given to students who use complete sentences to correctly answer all the questions and accurately complete the two data tables provided. Extra credit should be given to students who complete any of the follow-up activities.

www.birminghamzoo.com/ao/arthrop.htm (Animal Omnibus)
http//www.gene.com/ae/AE/AEC/AEF/index.html (Access Excellence)

1. From the list below, choose the class to which the copepod you studied in this activity belongs.

   Crustacea *Chilopoda* *Diplopoda* *Arachnida* *Insecta*

   Explain your answer.
2. The eyepiece of a microscope magnifies 10 times, and the objective magnifies 40 times. What is the total magnification of an object seen under this microscope?

Name____ Date ____________________

Under the Microscope: A View of Copepods

STUDENT ACTIVITY PAGE

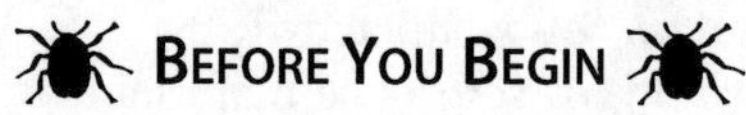

While many arthropods are easily visible to the naked eye, many others are microscopic. Today you will examine copepods, tiny arthropods that are barely visible without magnification. In order to get a better view of the *anatomy* and activities of these animals, you will observe them under a stereomicroscope.

Copepods may be found in either *marine* or freshwater. Most are *herbivores* and are a major food source for fish and other aquatic animals. In fact, they comprise most of the diet of some of the largest whales. The organisms that you will observe today live in freshwater. They have been shipped to your school in a jar that also contains a food supply of aquatic plants, or *algae.*

- Jar containing copepods in freshwater
- Plastic spoon or scoop
- Small culture dish
- Stereomicroscope
- Medicine dropper

PROCEDURE

You will be working with a partner with whom you will share your observations. Work as a team to answer the questions and to record your data and answers in the Data Collection and Analysis section.

1. Using only the naked eye, observe the copepods in their jar for one minute. You can recognize them as light beige moving organisms. Write your observations of the organism, including a physical description and a discussion of its locomotion and any other life functions that you see. Draw your copepod.
2. Using your spoon or scoop, remove from the jar a small amount of water containing copepods. Place it in the culture dish. If you do not see any copepods moving in the culture dish, return the water to the jar and try again.
3. Examine your stereomicroscope. You will see numbers on the *eyepiece* and *objective* lenses that indicate the number of times each lens magnifies the object (e.g., 10X). To obtain the total magnification, multiply the magnification of the eyepiece by the magnification of the objective.
4. Place the culture dish on the stage of the stereomicroscope. Remember that under a microscope, the image is upside down and backwards. Therefore, if your copepod swims out of the field of vision to the right, bring it back by moving the dish to the right. Always move your dish in the same direction that the organism moves.
5. Observe the copepod for one minute. What new information does the microscope reveal? What evidence do you have that the organism is an arthropod? Draw the organism. Label as many structures as possible.
6. Describe the copepod's swimming motion. Try to identify the structures involved in locomotion.

(continued)

© 1998 J. Weston Walch, Publisher

Name________________________________ Date ________________

Under the Microscope: A View of Copepods *(continued)*

STUDENT ACTIVITY PAGE

7. Refer to Table 1 in the first activity (Finding Arthropods: The Backyard Laboratory) or a library book. To which class do copepods belong? Why?
8. Gently place the tip of your medicine dropper in the water, near but not touching the copepod. How does the copepod react? Wait a few moments and try again. How might this type of behavior help the animal survive?
9. Hold your medicine dropper just above the surface of the water. Create a current by squeezing some air onto the surface. How does the copepod react? Do this again.
10. After completing your observations, return all copepods to the original jar. Clean up your lab station and thoroughly wash your hands. If time permits, discuss your results with a neighboring group.

DATA COLLECTION AND ANALYSIS

1. List your observations of the copepod in the table below

Description of Copepod

No Magnification	Magnified ______ 3

2. Read your description of the copepod to your partner. Based on your description, ask him or her to draw a copepod. Discuss any problems that your partner had in drawing from your description. Revise your description and try again.
3. In the space below, draw a picture of your copepod as seen under the microscope. Label as many structures as possible. Identify with the letter *L*, all structures used in locomotion.

(continued)

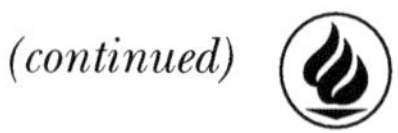

© 1998 J. Weston Walch, Publisher

Name____ Date ____________________

Under the Microscope: A View of Copepods *(continued)*

STUDENT ACTIVITY PAGE

4. What makes the copepod an arthropod? To which class does it belong? Why?
5. In the table below, list the copepod's reaction to stimuli.

Stimulus	Trial 1 Response	Trial 2 Response
Touch		
Water Current		

CONCLUDING QUESTIONS

Answer all questions on the back of the sheet or on a separate page.

1. What evidence do you have that the copepod is a *multicellular* rather than a *unicellular* organism?
2. Why was it necessary to repeat steps 8 and 9 before drawing any conclusions?
3. Another type of microscope used by biologists is the compound microscope. This type of microscope magnifies an object 100 to 400 times. Why wasn't this type of microscope used in this activity?

FOLLOW-UP ACTIVITIES

1. Use a textbook or the Internet to find out more about how copepods survive.
2. Design an experiment to answer the following question: What is the effect of __________ on copepods? Use variables such as light or changes in temperature. Be sure to treat your animals humanely.

Brine Shrimp: Determining the Optimum Saltwater Concentration

Teacher Resource Page

Instructional Objectives

Students will be able to

- mass objects on a triple-beam balance.
- record observations of living organisms.
- record data in a data table.
- draw conclusions based on data.

National Science Standards Addressed

Students demonstrate understanding of

- regulation and behavior.
- adaptation of organisms.
- big ideas and unifying concepts, such as cause and effect.

Students demonstrate scientific inquiry and problem-solving skills by

- identifying or controlling variables in experimental research settings.
- using evidence from reliable sources to develop explanations.
- evaluating the outcomes of investigations.
- working in teams to collect and share information and data.

Students demonstrate competence with the tools and technologies of science by

- using technology and tools to observe and measure objects, organisms, and phenomena.
- acquiring information from multiple sources, such as print, the Internet, and experimentation.

Students demonstrate effective scientific communication by

- representing data and results in multiple ways, such as tables, drawings, and writing.

Helpful Hints and Discussion

Time frame: Five single periods of instruction over the course of eight days
Structure: Groups of four students
Location: In class

The brine shrimp *Artemia salinas* is a small *crustacean* that lives in seawater but can also do well in water with salt concentrations between 1 percent and 30 percent. In this activity, students investigate the effect of different salt concentrations on the hatching of *Artemia salinas* eggs. Each group of four students should be assigned two different concentrations of saltwater. Make sure that each group has one concentration close to 3.5 percent (the concentration of seawater).

Brine shrimp are inexpensive and easy to maintain in the laboratory. The eggs may be purchased in a dehydrated, dormant condition from a pet store or a biological supply house such as Carolina Biological Supply Company. After being placed in an aerated saltwater tank at room temperature, the eggs will usually hatch within one to two days. Demineralized or spring water should be used to prepare the different concentrations of saltwater. An aquarium pump and air stone will provide oxygen, and a few grains of yeast will provide nourishment. The young brine shrimp, or *nauplii*, are *phototactic*. They can easily be collected by shining a light on one side of their tank.

The techniques that students learn in this activity may be used to design research projects to study the effects of other variables on this and other life processes. If students are planning to do future projects, make sure that they learn how to prepare different concentrations of salt water.

Materials

Each group of four students will need

- Triple-beam balance
- Weighing paper
- 130 grams of noniodized salt (NaCl)
- 1000-ml graduated cylinder
- Two glass or plastic one-liter tanks
- Wax marking pencil
- Two liters of spring or demineralized water
- Two grams of brine shrimp eggs
- Package of dry baking yeast
- Two aquarium pumps and air stones
- Flashlight or other light source
- Scoop
- Small finger bowl
- Stereomicroscope or hand lens

(continued)

ADAPTATIONS FOR HIGH AND LOW ACHIEVERS

High Achievers: These students should help lower achievers by working with them in a group. Also, encourage these students to perform additional experiments based on follow-up activity 3. Encourage high achievers in mathematics to complete follow-up activity 1.

Low Achievers: Provide a glossary and/or reference material for the italicized terms in this activity. Heterogeneous groupings will enable students of higher ability to assist those having difficulty carrying out the activity. Encourage these students to complete follow-up activity 2.

SCORING RUBRIC

Full credit should be given to students who use complete sentences to correctly answer all the questions and accurately complete the three data tables provided. Extra credit should be given to students who complete any of the follow-up activities.

www.birminghamzoo.com/ao/arthrop.htm (Animal Omnibus)

Base your answers to the questions on the paragraph below. To receive full credit, answer in complete sentences.

A student performs an experiment to determine the survival rate of brine shrimp at different salt concentrations. In tank A, he places 1 gram of brine shrimp eggs in 3.5 percent saltwater at 27° F. In tank B, he places 1 gram of brine shrimp eggs in 2.0 percent salt water at 30° F. After five days, he finds that there are 20 *nauplii* in tank A and 30 *nauplii* in tank B. He concludes that brine shrimp eggs have a better survival rate at 2.0 percent salt concentration.

1. Why was this experiment unscientific?
2. State one part of this experiment that was well done.
3. How should the experiment have been carried out?

Name____ Date ____________________

Brine Shrimp: Determining the Optimum Saltwater Concentration

STUDENT ACTIVITY PAGE

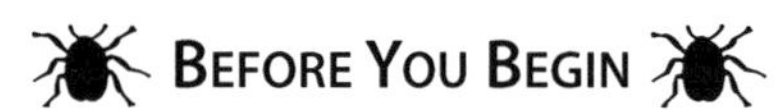

BEFORE YOU BEGIN

An important member of the food chain in *marine* environments is the brine shrimp, *Artemia salinas.* This small *crustacean* lives in seawater, which has a salt concentration of 3.5 percent. However, it also does well in water with salt concentrations ranging from 1 percent in brackish water to 30 percent in saltwater lakes that have no outlet. *Artemia* are *herbivores,* that eat *nonfilamentous algae* or yeast.

In this activity, you will investigate the effect of different salt concentrations on the hatching of *Artemia salinas* eggs. The techniques that you learn may be used to design research projects to study the effects of other variables on this and other life processes. Brine shrimp eggs can exist in a dormant condition if they are dehydrated. In fact, when you purchase them they are shipped to you in that state. After being placed in saltwater at room temperature, they will usually hatch within one to two days.

Brine shrimp are easy to maintain in the laboratory. They may be kept at room temperature in a small glass or plastic tank that will hold at least one liter of water. Demineralized or spring water should be used to prepare the different concentrations of saltwater. An aquarium pump and air stone will provide oxygen, and a few grains of yeast will provide nourishment. The young brine shrimp, or *nauplii,* are *phototactic.* They can easily be collected by shining a light on one side of their tank and scooping them up when they respond to the light.

MATERIALS

- Triple-beam balance
- Weighing paper
- 130 grams of noniodized salt (NaCl)
- 1000-ml graduated cylinder
- Two glass or plastic one-liter tanks
- Wax marking pencil
- Two liters of spring or demineralized water
- Two grams of brine shrimp eggs
- Package of dry baking yeast
- Two aquarium pumps and air stones
- Flashlight or other light source
- Scoop
- Small finger bowl
- Stereomicroscope or hand lens

PROCEDURE

1. *Preparation:* Your teacher will assign two solutions of saltwater, ranging from 1 to 6 percent, for your group to prepare. For each solution, mass the correct number of grams of salt and add it to one liter of spring or demineralized water. Stir until the salt is dissolved. Let it stand at room temperature for one day. Aerate the water with an aquarium pump and air stone.

Table 1: Preparation of 1 liter of Saltwater

%	NaCl (g)	%	NaCl (g)	%	NaCl (g)	%	NaCl (g)
1.0	10	1.5	15	2.0	20	2.5	25
3.0	30	3.5	35	4.0	40	4.5	45
5.0	50	5.5	55	6.0	60	6.5	65

(continued)

Name____ Date ______________________________

Brine Shrimp: Determining the Optimum Saltwater Concentration *(continued)*

STUDENT ACTIVITY PAGE

2. *Day 1:* Add one gram of brine shrimp eggs and a few grains of yeast to each tank. Continue to aerate the tank at room temperature for two more days.
3. *Day 3 (48 hours later):* Observe each tank for one minute. What evidence do you see that eggs have hatched? Write your observations in the Data Collection and Analysis section.
4. Shine a flashlight above one corner of the tank. After one minute, use the scoop to gather the shrimp *nauplii* that have congregated and place them in the finger bowl.
5. Count the number of *nauplii* in the bowl. Use a hand lens or stereomicroscope to help you make observations. Draw pictures of the organisms. Label as many parts as possible. Why are they classified as crustaceans?
6. Return your organisms to the original tank.
7. Repeat steps 4 through 6 with the organisms in your second tank.
8. *Day 5 (96 hours later):* Repeat steps 4 through 6 for both tanks.
9. Compare your results with those of other groups in your class.
10. Return all the organisms to the tank that has the most survivors.

DATA COLLECTION AND ANALYSIS

1. What evidence did you find that eggs hatched?

____

2. Complete the following table.

Table 2: Hatching of Brine Shrimp Eggs (After 48 hours)

Tank 1 **Salinity:** ________%	**Tank 2** **Salinity:** ________%
Observations of organisms in tank **Number of *nauplii*** **Drawing**	

(continued)

Name______________________________________ Date ______________________

Brine Shrimp: Determining the Optimum Saltwater Concentration *(continued)*

STUDENT ACTIVITY PAGE

3. Complete the following table:

Table 3: Hatching of Brine Shrimp Eggs (After 96 hours)

Tank 1 Salinity: ________%	Tank 2 Salinity: ________%
Observations of organisms in tank **Number of *nauplii*** **Drawing**	

4. Why are brine shrimp classified as crustaceans?

____

5. After comparing your data with that of the other groups in your class, complete the following table.

Table 4: Summary of Hatching of Brine Shrimp Eggs After 96 Hours

Salinity	Number of Nauplii

(continued)

© 1998 J. Weston Walch, Publisher

Name____________________________________ Date ______________________

Brine Shrimp: Determining the Optimum Saltwater Concentration *(continued)*

STUDENT ACTIVITY PAGE

CONCLUDING QUESTIONS

Answer all questions on the back of this sheet or on a separate page.

1. Why are brine shrimp eggs shipped in a dormant state?
2. What evidence do you have that the *nauplii* are *phototactic?*
3. Why are there likely to be fewer types of algae for the brine shrimp to eat in water that has a salt concentration of 30 percent than in water that has a salt concentration of 3.5 percent?

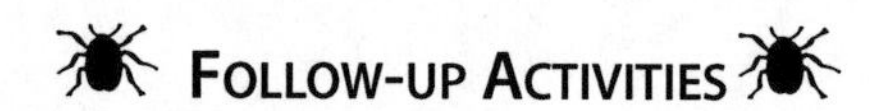

FOLLOW-UP ACTIVITIES

1. Refer to Table 1. Devise a mathematical formula to prepare 1 liter of saltwater at any concentration.
2. Use your textbook, other printed material, or the Internet to learn more about food chains in marine environments. Using words and/or pictures, prepare a report of an aquatic food web to present to your class.
3. Design an experiment to determine the effect of acid rain or another ecological variable on the hatching of *Artemia salinas.*

Terrestrial Crustaceans: Pill Bugs and Sow Bugs

TEACHER RESOURCE PAGE

INSTRUCTIONAL OBJECTIVES

Students will be able to

- record observations of living organisms.
- record data in a data table.
- perform simple calculations.
- formulate hypotheses about aspects of isopod behavior.
- draw conclusions based on data.

NATIONAL SCIENCE STANDARDS ADDRESSED

Students demonstrate understanding of

- structure and function in living systems.
- regulation and behavior, such as response to environmental stimuli.
- evolution, diversity, and adaptation of organisms, such as speciation, adaptation, and variation.
- big ideas and unifying concepts, such as form and function, and cause and effect.

Students demonstrate scientific inquiry and problem-solving skills by

- identifying variables in research settings.
- working in teams to collect and share information and ideas.

Sudents demonstrate competence with the tools and technologies of science by

- using technology and tools such as traditional laboratory equipment to observe and measure organisms and phenomena.
- collecting and analyzing data using concepts in mathematics.
- acquiring information from multiple sources, such as print, the Internet, and experimentation.

Students demonstrate effective scientific communication by

- arguing from evidence, such as data produced through his or her own experimentation.
- communicating in a form suited to the purpose and the audience, such as writing instructions that others can follow.

HELPFUL HINTS AND DISCUSSION

Time frame: Single period of instruction
Structure: Groups of four students
Location: In class

In this activity, students compare and contrast pill bugs and sow bugs. Both are terrestrial isopods that live in soil and eat decomposed plant material. Although they are called bugs, they are not insects but belong to the two groups of *crustaceans* that can spend their entire lives on land. They are similar in appearance but belong to different families. They can easily be distinguished from each other by their reactions when disturbed. Pill bugs will roll up into a ball or pill; sow bugs will not.

These animals are excellent laboratory animals because they are inexpensive (or free), easy to maintain, and require little care. They also move slowly and reproduce in large numbers. They may be kept in a large jar or bucket with a layer of 2 to 3 cm of moist soil on the bottom. Add slices of potato for food and small pieces of wood for the isopods to hide under. You may find them under rocks or purchase them from a biological supply house, such as Carolina Biological Supply Company.

When working with isopods, simply scoop them up with a plastic spoon or similar instrument and transfer them to the container in which you will be examining them. Although these animals are completely harmless to humans, some people may be allergic to certain arthropods. Therefore, it is advisable to wear gloves if you handle them. Remember to stress that these organisms are alive and should be treated humanely.

MATERIALS

Each group of four students will need

- Container of pill bugs
- Container of sow bugs
- Metric ruler
- Plastic gloves
- Petri dish
- Hand lens or stereomicroscope
- Deep tray or shoe box
- Plastic spoon or similar instrument
- Stopwatch or clock with a second hand

(continued)

Adaptations for High and Low Achievers

High Achievers: These students should help lower achievers by working with them in a group. Also, encourage these students to perform additional behavioral experiments based on follow-up activity 3.

Low Achievers: During your introduction, review characteristics of arthropods, particularly crustaceans and insects. Provide a glossary and/or reference material for the italicized terms in this activity. Heterogeneous groupings will enable students of higher ability to help those having difficulty carrying out the activity. Encourage these students to complete follow-up activity 2.

Scoring Rubric

Full credit should be given to students who use complete sentences to correctly answer all the questions and accurately complete the two data tables provided. Extra credit should be given to students who complete any of the follow-up activities.

http://www.gene.com/ae/AE/AEC/AEF/index.html (Access Excellence)
http://www.weneedyou.com./clark_bugs/sowpillbug.html (Sow Bug/Pill Bug)

1. A student brings a jar containing terrestrial isopods to class. Describe two ways that you can determine whether they are pill bugs or sow bugs.
2. Why are these terrestrial isopods believed to be more closely related to shrimps and lobsters than to beetles or ladybugs?

Name______________________________ Date ______________

Terrestrial Crustaceans: Pill Bugs and Sow Bugs

STUDENT ACTIVITY PAGE

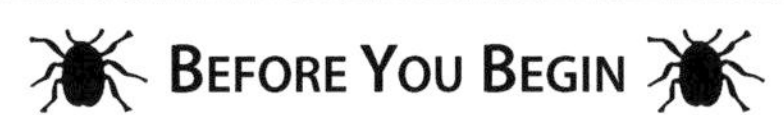

BEFORE YOU BEGIN

Neither pill bugs nor sow bugs are bugs. They belong to the two groups of *crustaceans* that can spend their entire lives on land. Both of these *terrestrial isopods* live in soil and eat decomposed plant material. Although they are similar in appearance, they can easily be distinguished from each other by their reactions when disturbed. Pill bugs will roll up into a ball or pill; sow bugs will not.

These animals are easy to maintain in the laboratory and require little care. They may be kept in a large jar or bucket with a 2- to 3-cm layer of moist soil on the bottom. Add slices of potato for food and small pieces of wood for the isopods to hide under.

In this activity, you will learn about similarities and differences between pill bugs and sow bugs. They move rather slowly and are easy to handle. Simply scoop them up with a plastic spoon or similar instrument and transfer them to the container in which you will be examining them. Although these animals are completely harmless to humans, some people may be allergic to certain arthropods. Therefore, it is advisable to wear gloves if you handle them. Remember that these organisms are alive and should be treated humanely. At the end of the activity, return them to their original container.

MATERIALS

- Container of pill bugs
- Container of sow bugs
- Metric ruler
- Plastic gloves
- Petri dish
- Hand lens or stereomicroscope
- Deep tray or shoe box
- Plastic spoon or similar instrument
- Stopwatch or clock with a second hand

PROCEDURE

1. Observe pill bugs and sow bugs in their containers for two minutes. Write a description of each type of animal in the Data Collection and Analysis section. Be specific enough in your description that the members of your learning group can distinguish between the two organisms when they are placed together.
2. Transfer three pill bugs to a petri dish. Cover the dish. Using a hand lens or stereomicroscope, count the number of antennae, body segments, and legs on each pill bug. It might help to hold the dish up over your head so that you can see the organisms' *ventral* surfaces. Record this and all observations in the appropriate tables in the Data Collection and Analysis section.
3. Transfer the pill bugs to a deep tray or shoe box. Touch one gently with the back of your spoon. How does it react?
4. After the animal recovers, touch the other bugs in a similar manner. This time use a stopwatch or a clock with a second hand to determine the length of time it takes each pill bug to recover.

(continued)

© 1998 J. Weston Walch, Publisher

Name________________________________ Date ________________

Terrestrial Crustaceans: Pill Bugs and Sow Bugs *(continued)*

STUDENT ACTIVITY PAGE

5. Measure the length and width of each pill bug, in mm.
6. Place one of the pill bugs on its back, or *dorsal* surface. Place the top of your spoon along the midline of the *ventral* surface. Try to lift the pill bug. How does it react? Do the same to the other two pill bugs. How do they react?
7. Measure the length of your tray or box in centimeters. Place an animal at one end. How long does it take it to get to the other end? Calculate the speed in cm/sec. Do the same with each of the other pill bugs.
8. Return the pill bugs to their original container. Repeat steps 2 to 7 with sow bugs.
9. When you have finished, return all organisms to their original containers, clean up your laboratory station, and wash your hands thoroughly.

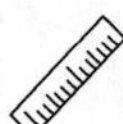

DATA COLLECTION AND ANALYSIS

1. Complete the following table.

Table 1: Description of Terrestrial Isopods

Pill Bug	Sow Bug

2. Complete the following table.

Table 2: Characteristics of Terrestrial Isopods

Characteristic	Pill bug	Sow Bug
# of antennae		
# of body segments		
Pairs of legs/segment		
Reaction to touch		
Average recovery time (sec)		
Average length (mm)		
Average width (mm)		
Ratio of length/width		
Can it be lifted from ventral surface?		
Average speed (cm/sec)		

(continued)

Name________________________________ Date ________________

Terrestrial Crustaceans: Pill Bugs and Sow Bugs *(continued)*

STUDENT ACTIVITY PAGE

3. Based on the information in Table 2, write a paragraph or prepare a concept map on the back of this sheet describing the two types of terrestrial isopods (pill bugs and sow bugs) in as many ways as possible.

CONCLUDING QUESTIONS

Answer all questions on the back of this sheet or on a separate page.

1. How are these isopods adapted to survive in soil?
2. How is each type of isopod adapted to avoid enemies?
3. Why are pill bugs and sow bugs good organisms to study in a school laboratory?
4. Why are these organisms classified as crustaceans, rather than insects?
5. Why is isopod a good name for these organisms?

FOLLOW-UP ACTIVITIES

1. Use printed material or the Internet to learn about the anatomy and life cycle of other crustaceans, such as lobsters and shrimps. Present your findings as a written report or artwork.
2. Use printed material or the Internet to learn about food webs in the soil. Draw a diagram of a food web that includes isopods.
3. Design an experiment to learn more about the behavior of isopods. (For example: Can they learn? How strong is a pill bug?)

Thousand Legs: The Life of Millipedes

TEACHER RESOURCE PAGE

INSTRUCTIONAL OBJECTIVES

Students will be able to

- record observations of living organisms.
- draw and label organisms.
- perform simple calculations.
- record data in a data table.
- draw conclusions based on data.

NATIONAL SCIENCE STANDARDS ADDRESSED

Students demonstrate understanding of

- structure and function in living systems.
- regulation and behavior.
- big ideas and unifying concepts, such as order and organization and form and function.

Students demonstrate scientific inquiry and problem-solving skills by

- evaluating the outcomes of investigations.
- working in teams to collect and share information and ideas.

Students demonstrate competence with the tools and technologies of science by

- using technology and tools to observe and measure organisms.
- collecting and analyzing data using mathematical concepts, such as mean.

Students demonstrate effective scientific communication by

- representing data and results in multiple ways, such as tables, drawings, and writing.
- explaining concepts to other students.

MATERIALS

Each group of four students will need

- Tank containing at least 10 millipedes
- Plastic vial covered with foam plug
- Plastic gloves
- Metric ruler
- Plastic or cardboard shoe box
- Petri dish
- Piece of cardboard (10 cm wide X 5 cm high)
- Stereomicroscope
- Hand lens
- Piece of apple or potato

HELPFUL HINTS AND DISCUSSION

Time frame: Single period of instruction
Structure: Groups of four students
Location: In class

In this activity, students will examine adaptations of millipedes for locomotion, regulation, and nutrition. These slow-moving soil organisms are *herbivores* that feed on decayed plant material.

Millipedes may be maintained in the laboratory in a tank that contains a layer at least 6-cm thick of humus, or soil with decayed plant material. Pieces of egg carton will provide dark areas under which the organisms can burrow, and an occasional piece of apple or potato will provide moisture. Although the millipedes are not likely to climb up the sides of the tank, it is advisable to cover the tank with a screen.

When introducing the activity, demonstrate proper techniques for transferring the animals. They may be captured in a plastic vial, which should then be covered with a foam plug, or simply be picked up by hand. Remind your students that the animals are alive and should be treated humanely.

(continued)

Adaptations for High and Low Achievers

High Achievers: These students should help lower achievers by working with them in a group. Also, encourage these students to perform behavioral experiments based on follow-up activity 2.

Low Achievers: Provide a glossary and/or reference material for the italicized terms in this activity. Take time to review vocabulary, paying particular attention to prefixes such as *centi-* and *milli-*. Heterogeneous groupings will enable students of higher ability to assist those having difficulty carrying out the activity. Encourage these students to do follow-up activity number 3.

Scoring Rubric

Full credit should be given to students who use complete sentences to correctly answer all the questions and accurately complete the two data tables provided. Extra credit should be given to students who complete any of the follow-up activities.

Internet Tie-ins

http://azstarnet.com/~sasi. (Arthropod World)
www.birminghamzoo.com/ao/arthrop.htm (Animal Omnibus)
http//www.gene.com/ae/AE/AEC/AEF/index.html (Access Excellence)

Quiz

Base your answers to the following questions on the paragraph below and your knowledge of biology.

Millipedes, like insects and spiders, belong to the phylum *Arthropoda.* However, the differences between millipedes and insects are sufficient to place the millipedes in their own class, *Diplopoda.*

1. Why are millipedes placed in the phylum *Arthropoda?*
2. List two differences between organisms in the class *Insecta* and those in the class *Diplopoda.*

Name ______________________ Date ______________________

Thousand Legs: The Life of Millipedes

STUDENT ACTIVITY PAGE

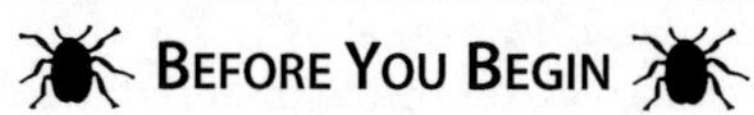

BEFORE YOU BEGIN

Millipedes are among the more interesting arthropods that burrow in soil. While they may appear to have thousands of legs, this is not the case. Close examination reveals that the legs are easily counted and that each segment has the same number of legs. Millipedes are *herbivores* that feed on decayed plant material.

Millipedes may be maintained in the laboratory in a tank that contains a layer of humus at least 6 cm thick. Pieces of egg carton will provide dark areas under which the organisms can burrow, and an occasional piece of apple or potato will provide moisture. Although the millipedes are not likely to climb up the sides of the tank, it is advisable to cover the tank with a screen.

In this activity, you will explore some of the adaptations and life functions of millipedes. Although these animals are not harmful to humans, some people may be allergic to various arthropods. Therefore, it is advisable to wear plastic gloves when handling them. Millipedes may be transferred by capturing them in a plastic vial, which should then be covered with a foam plug. However, it is just as easy to gently pick them up in your hands. Remember that they are alive and should be treated humanely. Handle them gently and do not make any sudden noises or otherwise startle them. When you have completed this activity, return the millipedes to their tank.

MATERIALS

- Tank containing at least 10 millipedes
- Plastic vial covered with foam plug
- Plastic gloves
- Metric ruler
- Plastic or cardboard shoe box
- Petri dish
- Piece of cardboard (10 cm wide X 5 cm high)
- Stereomicroscope
- Hand lens
- Piece of apple or potato

PROCEDURE

1. Observe millipedes in the tank for two minutes. In the Data Collection and Analysis section, write a description of a millipede and draw a picture of it. Label as many parts as possible. Be precise.
2. Remove three millipedes, one at a time, from their tank. Measure each to the nearest cm. Count the number of segments on each millipede. Using the following formula, determine the average length of the segments: Average length of segment = total length (cm) / number of segments. Return two of the millipedes to their tank after measuring and counting their segments.
3. Place the other millipede in the shoe box. Observe its motion, including the order in which the legs move. Does it move in a straight or curved path? Which legs seem to move first? Last? Do the legs in the same segment move at the same time? In addition to legs, are any other structures involved in motion? If so, which one(s)? Write your answers in the Data Collection and Analysis section.
4. Place the millipede in a petri dish. Cover the dish and hold it over your head so that you can observe the *ventral* surface without turning the dish upside down. What additional clues about the motion of millipedes can you now observe?

(continued)

Name____ Date ____________________

Thousand Legs: The Life of Millipedes *(continued)*

STUDENT ACTIVITY PAGE

5. Return the millipede to the shoe box. Create an obstacle by holding the piece of cardboard upright in the center of the box. Leave room for a passageway at both sides. How does the millipede detect the obstacle? What is its reaction?
6. Repeat step 5 with another millipede. Note similarities and differences between the behavior of the two animals.
7. Place the millipede upside down on its *dorsal* surface on the tabletop. Describe how it rights itself. Make your description as precise as possible. Try it again. Does the millipede react in the same way?
8. Repeat step 7 with another millipede. Note similarities and differences between the behavior of the two animals.
9. Place a millipede in the petri dish and cover it. Using your hand lens and/or stereomicroscope, examine the mouth of the animal. How is it adapted to obtain nutrients from decayed plant material in the soil?
10. Place a piece of apple or potato in the dish with the millipede. If possible, observe how the animal obtains moisture. What structures in addition to the mouth are involved in obtaining moisture?
11. When you have completed this activity, be sure that all millipedes have been returned to their tank and clean your laboratory station. Wash your hands thoroughly.

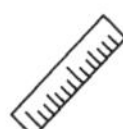

DATA COLLECTION AND ANALYSIS

1. In the space below, write a description of your millipede and draw a picture of the animal. Label as many parts as possible.

2. Complete the table below.

Table 1: Length of Millipedes

Millipede #	Length (cm)	# of Segments	Average Length of Each Segment (cm)
1			
2			
3			

(continued)

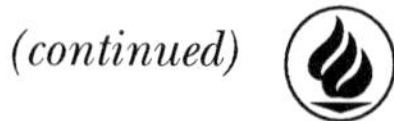

© 1998 J. Weston Walch, Publisher

Name________________________________ Date ________________

Thousand Legs: The Life of Millipedes *(continued)*

STUDENT ACTIVITY PAGE

3. In the table below, describe each type of behavior in millipedes 1 and 2.

Table 2: Reactions of Millipedes

Type of Behavior	Millipede 1	Millipede 2
Locomotion (path, which legs move first, how the legs of the same segment move, other structures involved)		
Reaction to obstacles		
Righting behavior		

4. How is the millipede adapted to obtain nutrients from decayed plant material in soil? Be as specific as possible.

CONCLUDING QUESTIONS

Answer all questions on the back of this sheet or on a separate page.

1. Refer to question 1 in the Data Collection and Analysis section. Read your description of a millipede to your teammates. Have them draw a picture of a millipede from your description. Ask them for suggestions about how to improve your written description. Based on their suggestions, revise your description.
2. Why are millipedes classified as arthropods rather than segmented worms?
3. Millipedes belong to the class of arthropods called *Diplopoda.* Why is this a good name? (*Hint:* Use your dictionary to find words beginning with *diplo-* and *pod-*.)

FOLLOW-UP ACTIVITIES

1. At first glance, *centipedes* may sometimes be confused with millipedes. Use your textbook, other printed material, or the Internet to learn about centipedes. Write about their similarities and differences, including the way they obtain nutrition. Why aren't centipedes good organisms to use in a laboratory activity?
2. Design an experiment to discover whether millipedes can learn. (For example, will they learn to follow a certain path in a Y-maze in order to get food?)
3. Use your dictionary to learn the meanings of each of the following prefixes: *hex-, dec-, centi-, milli-, uni-, multi-, poly-, bi-, tri-, pseudo-, cephalo-, gastro-, pod-*. Make up a name for an imaginary animal, using at least two names from the list. Draw a picture of your animal.

© 1998 J. Weston Walch, Publisher

The Web of Death: A Spider's Trap

TEACHER RESOURCE PAGE

INSTRUCTIONAL OBJECTIVES

Students will be able to

- record observations of a living organism.
- perform measurements of length, using the metric system.
- record data in a data table.
- perform simple calculations.
- draw conclusions based on data.
- make a labeled drawing of a spider.

NATIONAL SCIENCE STANDARDS ADDRESSED

Students demonstrate understanding of

- populations and ecosystems, such as the roles of predators and prey.
- adaptations of organisms.
- big ideas and unifying concepts, such as form and function.

Students demonstrate scientific inquiry and problem-solving skills by

- identifying variables.
- working in teams to collect and share information and ideas.
- evaluating the accuracy and outcomes of investigations.

Students demonstrate competence with the tools and technologies of science by

- using tools to observe and measure objects and organisms.
- analyzing data using concepts and techniques in Mathematics Standards.
- acquiring information from multiple sources.

Students demonstrate effective scientific communication by

- Recording data in a variety of formats.

MATERIALS

Each pair of students will need

- Hand lens
- Metric ruler
- Two cotton swabs
- Plastic gloves
- Laboratory notebook, pen, and pencil

HELPFUL HINTS AND DISCUSSION

Time frame: Single period of instruction
Structure: Pairs of students
Location: At home or outdoors

In this activity, students explore spiders and their webs to learn how spiders obtain food. Webs may be found indoors or outdoors almost anywhere there are two surfaces. Because one can't always predict where a spider's web will be found, this activity cannot easily be done in class.

When introducing this activity, stress the following.

- The behavior of organisms is often unpredictable. Therefore, observing, collecting results, and drawing conclusions about organisms in their habitats require patience and an open mind.
- Sudden movements or loud noises will interfere with experiments.
- Organisms are living and should be treated humanely.
- All spiders are predators.
- Although most spiders will not harm humans, they should not be touched.
- Always work with a partner or take someone with you when looking for spiderwebs.
- Wear plastic gloves when touching webs to avoid allergic reactions.
- Wash hands thoroughly after the experiment.

(continued)

Adaptations for High and Low Achievers

High Achievers: These students should help lower achievers by working with them in a group. Also, encourage these students to perform additional experiments based on follow-up activities 1, 2, or 3.

Low Achievers: Provide a glossary and/or reference material for the italicized terms in this activity. Heterogeneous groupings will enable students of higher ability to assist those having difficulty carrying out the activity. Take time to introduce and practice the mathematical concepts.

Scoring Rubric

Full credit should be given to students who use complete sentences to correctly answer all the questions and accurately complete the data tables. Extra credit should be given to students who complete any of the follow-up activities.

Internet Tie-ins

http://biology.arizona.edu (collecting, keeping, and caring for spiders)
http://azstarnet.com/~sasi.aw.html (Arthropod World)
www.birminghamzoo.com/ao/arthrop.htm (Animal Omnibus)

Quiz

1. Using the information you have collected during this activity and each of the words below, draw a concept map and write a short essay explaining how spiders get food.

   spider insect prey predator web sticky nonsticky

2. What evidence do you have that a spider is an arachnid and not an insect?

Name________________________________ Date ______________________

The Web of Death: A Spider's Trap

STUDENT ACTIVITY PAGE

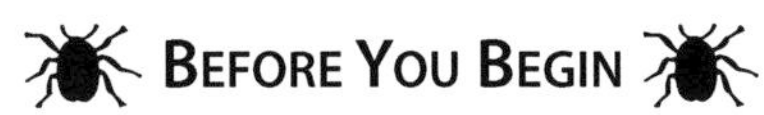

BEFORE YOU BEGIN

Today you will explore the way spiders obtain their food. Spiders are arthropods that belong to the class *Arachnida.* They have two body segments, the *cephalothorax* and the *abdomen,* which are joined by a narrow waist. They also have four pairs of legs and usually eight simple eyes. They do not have antennae. All spiders are *predators*; most build webs to capture their *prey.*

In this activity, you will observe and discover some characteristics of spiders and their webs. Webs can be found almost anywhere there are two surfaces. Outdoors you may find them between branches of trees, on the sides of buildings, or on park benches. Indoors, look for webs in the corners of rooms or hanging from lamps or other pieces of furniture that have not recently been dusted.

Because you can't always predict where you will find a spider's web, this activity cannot easily be done in class. When looking for spiderwebs, be careful. Although most spiders will not harm humans, do not touch them. Work with a partner or take someone with you. Wear gloves when touching spiderwebs. Make sure that you wash your hands thoroughly after this laboratory activity.

MATERIALS

Each pair of students will need

- Hand lens
- Metric ruler
- Two cotton swabs
- Plastic gloves
- Laboratory notebook, pen, and pencil

PROCEDURE

You will work with a partner. This will enable you to share your observations. Work as a team to answer the questions and to record your data and answers in your laboratory notebook and on the charts.

1. Find a spider's web. Observe it with both the naked eye and a hand lens. Take notes on your observations. Look for the spider. Where is it? What evidence do you have that it is an arachnid? Draw a picture of your spider. Label as many parts as possible and indicate the color of the organism.
2. Look for the prey. How many are there? Are they all insects? How do you know?
3. Draw a picture of the web. Indicate the location of the spider and the prey.
4. Measure the length and width of the web to the nearest cm. Calculate the area using the following formula: area = length X width. Your answer will be in square centimeters.
5. Using a cotton swab, gently touch various parts of the web, including the center, a horizontal strand, a vertical strand, an area where prey is, and the area where the spider is. Note the spider's reaction as you touch each part of the web. Are there differences in stickiness? If so, explain how each area differs. Examine the sticky and nonsticky parts with a hand lens. What do you notice?

(continued)

Name____ Date ____________________________

The Web of Death: A Spider's Trap *(continued)*

STUDENT ACTIVITY PAGE

6. With your partner, locate a *dragline.* This is a vertical strand near the outside of the web, along which the spider may travel.
   (a) Measure the strand to the nearest 0.1 cm. A = __________cm
   (b) Gently pull the strand without breaking it. While one partner holds the strand, have the other partner measure it to the nearest 0.1 cm. B = __________cm
   (c) Let go of the strand. Remeasure it to the nearest 0.1 cm. C = __________cm
   (d) You can now perform the following calculations about the web:
   increase in size when stretched = $B - A$
   percent of increase when stretched = $[(B - A)/A] \times 100$
   final change = $C - A$
   Percent of Final Change = $[(C - A)/A] \times 100$
7. Find a second web, preferably from a different type of spider. Using this web, repeat steps 1 through 6.
8. After completing your observations and recording your data, wash your hands thoroughly.

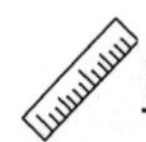

DATA COLLECTION AND ANALYSIS

1. Record your observations of the spider and its web. Before submitting your observations to your teacher, revise your writing, making sure that you have answered in complete sentences and checked your spelling. Draw and label your spiders in the space below.

Spider 1	Spider 2

2. Record your observations of the prey in the table below. How can you tell whether the prey are insects?

	Web 1	Web 2
# of insects		
# of noninsects		

3. Draw a picture of each web on the back of this sheet. Indicate the location of spider and prey.

(continued)

Name________________________________ Date ____________________

The Web of Death: A Spider's Trap *(continued)*

STUDENT ACTIVITY PAGE

4. Indicate the measurements of each web in the table below. Include the proper units (cm or square cm) in your answer.

	Web 1	Web 2
Length		
Width		
Area (L 3 W)		

5. In the table below, tell whether each area was *sticky* or *nonsticky* when touched with the cotton swab. Tell what the spider did when you touched each area.

	Web 1	Web 2
Center		
Horizontal strand		
Vertical strand		
Prey area		
Spider area		

On the back of this sheet, tell how the sticky area differed in appearance from the nonsticky area.

6. In the table below, complete the calculations you made on one strand of the web.

	Web 1	Web 2
Increase when stretched		
% increase		
Final change		
% change		

(continued)

Name________________________________ Date ________________

The Web of Death: A Spider's Trap *(continued)*

STUDENT ACTIVITY PAGE

CONCLUDING QUESTIONS

Answer all questions on the back of this sheet or on a separate page.

1. Why would a scientist want to perform this experiment more than once before drawing any conclusions?
2. List as many similarities and differences between the two webs as you can.
3. Based on your calculations in step 6, explain if the dragline is *elastic.*

FOLLOW-UP ACTIVITIES

1. Find out about *orb webs, sheet webs,* and *dome webs.* How do they compare with each other? Based on your research, tell what type(s) of webs you examined in this investigation.
2. Using books on arthropods, identify the type(s) of spiders and prey you saw.
3. Calculate the distances between prey in each web. Note similarities and differences.

From Egg to Adult: A Fruit Fly's Life

Teacher Resource Page

Instructional Objectives

Students will be able to

- record observations of a living organism.
- list stages in the life cycle of an insect.
- draw and label stages of *Drosophila's* life cycle.
- formulate hypotheses.
- draw conclusions based on data.

National Science Standards Addressed

Students demonstrate understanding of

- reproduction and heredity.
- big ideas and unifying concepts, such as form and function.

Students demonstrate scientific inquiry and problem-solving skills by

- identifying variables in experimental and nonexperimental research conditions.
- using evidence from reliable sources to develop descriptions and explanations.
- working in teams to collect and share information and ideas.

Students demonstrate competence with the tools and technology of science by

- acquiring information from multiple sources, such as print, the Internet, and experimentation.

Students demonstrate effective scientific communication by

- representing data and results in multiple ways, such as drawings and technical writing.
- explaining a scientific concept to other students.

Materials

Each group of four students will need

- Diluted disinfectant, such as Lysol™, in a wash bottle
- Paper towels
- Vial of vestigial-winged (vg) culture of *Drosophila melanogaster*
- Instant *Drosophila* Medium (blue)
- Several empty *Drosophila* culture vials, foam plugs, and vial caps
- Scoop or tablespoon
- Water
- *Drosophila* sorting brush or thin paintbrush
- Hand lens
- Stereomicroscope
- Clean probe
- Glass slide

(continued)

Helpful Hints and Discussion

Time frame: Single period of instruction
Structure: Groups of four students
Location: In class

The fruit fly, *Drosophila melanogaster*, undergoes complete metamorphosis as it develops from egg to adult. In this activity, your students make detailed observations of the changes that occur as the life cycle progresses. They also become familiar with techniques involved in working with this animal and prepare to perform simple controlled experiments using *Drosophila* as a research organism.

Drosophila are inexpensive and relatively easy to maintain in the laboratory. They are kept in plastic vials covered with foam plugs. The food at the bottom of the vial contains equal amounts of Instant *Drosophila* Medium and water. In order to make the white larvae more visible, it is suggested that you order blue medium.

One of the biggest problems encountered in working with this organism is mold. To prevent this, clean your tabletop and probes with Lysol or a similar disinfectant and add a few grains of yeast to the top of the food after it hardens. Make sure that your students wash their hands before beginning this activity. Keep all books and other materials away from the vials.

Because flies with vestigial wings cannot fly and move rather slowly, they are ideal for this experiment. However, you might want to refrigerate the flies for at least 15 minutes before beginning this activity. This will not harm the flies and for a short time will slow them down even more.

Adaptations for High and Low Achievers

High Achievers: These students should help lower achievers by working with them in a group. Also, encourage these students to perform additional experiments based on follow-up activities 2 and 3.

Low Achievers: Provide a glossary and/or reference material for the italicized terms in this activity. Heterogeneous groupings will enable students of higher ability to assist those having difficulty carrying out the activity. Encourage these students to perform additional experiments based on follow-up activities 1 and 4.

Scoring Rubric

Full credit should be given to students who use complete sentences to correctly answer all the questions and accurately complete the data table provided. Extra credit should be given to students who complete any of the follow-up activities.

Internet Tie-ins
www.carolina.com (Carolina Biological Supply Co.)
www.birminghamzoo.com/ao/arthrop/htm (Animal Omnibus)

Quiz Answer the questions below, based on the following paragraph and your work with *Drosophila melanogaster.*

The common fruit fly, *Drosophila melanogaster*, is an excellent laboratory organism that is used by scientists to study genetics. The flies have a short life cycle, and are inexpensive and easy to maintain in the laboratory. They also produce many offspring. One female may lay more than 200 eggs in its lifetime.

1. List the stages in the life cycle of the fruit fly, starting with the egg. Describe the appearance of the organisms at each stage.
2. During which stage do fruit flies eat the most? Why?
3. Although *Drosophila* are easy to maintain, students often have difficulty working with them. List one problem that you had when working with this organism and describe how you solved the problem or how you would solve it next time.

Name________________________________ Date ____________________

From Egg to Adult: A Fruit Fly's Life

STUDENT ACTIVITY PAGE

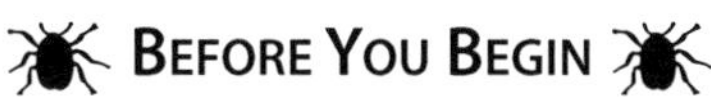

The fruit fly, *Drosophila melanogaster*, has a life cycle similar to the butterfly's. It is called complete *metamorphosis* and has four stages. The first is an egg. This develops into a larva (caterpillar stage), which at first glance looks like a small, white worm. After several days, the larva becomes a pupa, or cocoon. When they first form, the pupae resemble small, motionless grains of brown rice. Finally, an adult fly emerges. The entire life cycle takes approximately two weeks.

In this activity, you will study the characteristics of all four stages. You will use flies with *vestigial* (very small) wings because they cannot fly and are easier to handle.

When working with fruit flies, there are a few precautions you should take. The flies are small and fragile, so be careful with them. When transferring adults from one bottle (vial) to another, make sure that there is no space between the vials. This will prevent the flies from escaping. When transferring larvae, make sure that you work on a clean surface. If you first clean your tabletop and probes with a disinfectant, such as Lysol™, you will reduce the chances of picking up mold spores, which could destroy your culture.

The flies that you will observe have been kept in a vial that is covered with a sponge plug. This allows air to enter and leave. The food at the bottom of the vial contains equal amounts of Instant *Drosophila* Medium and water. A few grains of yeast have been added to help prevent mold from growing. The flies can survive in this container for several weeks.

Materials

- Diluted disinfectant, such as Lysol, in a wash bottle
- Paper towels
- Vial of vestigial-winged (vg) culture of *Drosophila melanogaster*
- Instant *Drosophila* Medium (blue)
- Several empty *Drosophila* culture vials, foam plugs, and vial caps
- Scoop or tablespoon
- Water
- *Drosophila* sorting brush or thin paintbrush
- Hand lens
- Stereomicroscope
- Clean probe
- Glass slide

Procedure

1. Observe the organisms in your vial for two minutes. Take notes on your observations. Draw as many stages as you see.

(continued)

Name________________________________ Date ________________

From Egg to Adult: A Fruit Fly's Life *(continued)*

STUDENT ACTIVITY PAGE

2. Prepare one vial of food in the following way: Into an empty vial, place one scoop or tablespoon of dry food. Add an equal amount of water. Swirl the contents gently to mix. When the mixture hardens (approximately one minute), add two or three grains of yeast. This will prevent mold from forming. Immediately cover the vial with a foam plug.
3. *Read all directions in step 3 before continuing.* Practice several times with two empty vials before trying to transfer flies.
   Transfer some vestigial flies to an empty vial by the following method:
   (a) Remove the foam plug from the empty vial and turn the vial upside down.
   (b) Take the vial that contains the flies. Gently tap the bottom against your tabletop. This should cause the flies to fall to the surface of the food.
   (c) Place the empty vial in one hand. Very quickly remove the foam plug from the vial that contains the fruit flies and place the empty inverted vial on top of it.
   (d) After the adult flies have moved into the empty vial, quickly replace both foam plugs, with the help of the other members of your group.
4. Look at the vial that contains the adult flies. Using Figure 1, try to distinguish between males and females. If necessary, use a hand lens to make observations.

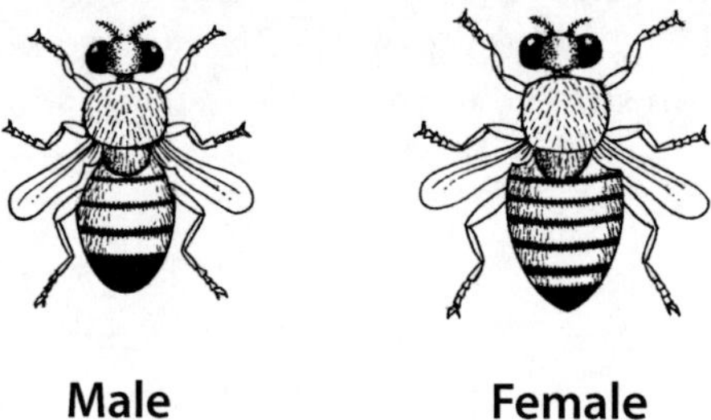

Male **Female**

Figure 1: Male and Female Vestigial *Drosophila melanogaster*[1]

5. Observe the organisms that remain on the sides of the original vial and in the food. Look for tunnels in the food. Use your hand lens or stereomicroscope to discover the cause of the tunnels. *Note:* The vial may be placed on its side on the stage of the microscope. Focus up and down until the tunnels come into view.
6. Examine the organisms on the sides of the vial, first with the naked eye and then with the hand lens or stereomicroscope. How do these compare with the organisms that were in the food?
7. Using your probe, carefully remove one larva from your vial and place it on a glass slide. Observe it under the stereomicroscope. Look for the digestive system. What color will it be? Why?
8. Locate a light brown pupa. Using your stereomicroscope, look for evidence of body parts, such as eyes, wings, or legs.
9. Repeat step 7, using the darkest pupa you can find. Based on your observations, explain which seems to be older.

[1] *Carolina Drosophila Manual.* (Burlington, North Carolina: Carolina Biological Supply Co., 1988)

(continued)

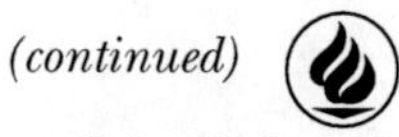

Name________________________________ Date ________________________

From Egg to Adult: A Fruit Fly's Life *(continued)*

STUDENT ACTIVITY PAGE

10. The eggs are microscopic. In which portion of the vial do you think they will be found? Why? Test your hypothesis.

DATA COLLECTION AND ANALYSIS

1. In the diagram of the vial below, draw and label the stages found at each location in the vial.

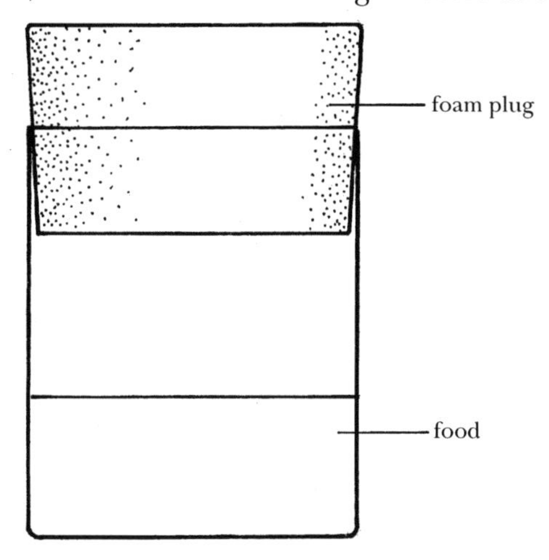

2. Complete the following chart. Label as many structures as possible in your drawings.

Life Cycle of *Drosophilia Melanogaster*

Stage	Drawing (Actual Size)	Drawing (As Seen Under Stereomicroscope)
Egg		
Larva		
Pupa		
Adult		

3. How were the tunnels in the food produced?

(continued)

© 1998 J. Weston Walch, Publisher

Name________________________________ Date ______________________

From Egg to Adult: A Fruit Fly's Life *(continued)*

Student Activity Page

4. Draw a picture of a larva as seen under the microscope. Indicate the location of its digestive system. What evidence indicates that the larva is an arthropod, rather than a worm?
5. In the space below, draw the light and dark pupae as seen under the microscope. Label as many structures as possible.

Light Pupa **Dark Pupa**

6. Which of the two pupae is older? How do you know?

Concluding Questions

Answer all questions on the back of this sheet or on a separate page.

1. Where did you expect the most eggs to be found? How did you test your hypothesis? Was your hypothesis correct?
2. Why do you think that insect pests such as clothes moths, cause more damage as larvae than as pupae?

Follow-up Activities

1. Based on the data above, predict what will happen during the next two weeks in the vial that contains only adult flies. Be very specific.
2. A *limiting factor* is an environmental condition (light, food, water, temperature, and so on) that will limit the size of a population and eventually cause its death. Based on this activity, what limiting factor do you think will eventually cause the death of your flies? Why?
3. *Drosophila melanogaster* are often used in experiments to study genetics, or heredity. Use texts, journals, or the Internet to learn about the genetics of *Drosophila.* Prepare a report to present to your class.
4. Prepare a poster on the life cycle of insects. Include at least two types of insects.

© 1998 J. Weston Walch, Publisher

UV Light: Effect on Hatching Time

TEACHER RESOURCE PAGE

INSTRUCTIONAL OBJECTIVES

Students will be able to

- record observations of a living organism.
- conduct controlled experiments.
- record data in a data table.
- perform simple calculations.
- draw conclusions based on data.

NATIONAL SCIENCE STANDARDS ADDRESSED

Students demonstrate understanding of

- structure and function in living systems.
- reproduction and heredity.
- big ideas and unifying concepts, such as form and function and cause and effect.
- health and disease, such as effects of drugs and toxic substances.

Students demonstrate scientific inquiry and problem-solving skills by

- identifying or controlling variables in experimental research settings.
- using evidence from reliable sources to develop descriptions, explanations, and models.
- working in teams to collect and share information and ideas.

Students demonstrate competence with the tools and technologies of science by

- using technology and tools to observe and measure organisms and phenomena.
- recording and storing data.
- collecting and analyzing data using concepts and techniques in Mathematics Standards.
- acquiring information from multiple sources, such as print, the Internet, and experimentation.
- recognizing sources of bias in data.

Students demonstrate effective scientific communication by

- representing data and results in multiple ways, such as numbers, tables, and graphs.
- arguing from evidence, such as data produced through their own experimentation.
- explaining a scientific concept or procedure to other students.

MATERIALS

Each group of four students will need

- Diluted disinfectant, such as Lysol, in a wash bottle
- Paper towels
- Clean probe
- Scoopula
- Three empty *Drosophila* culture vials, foam plugs, and vial caps
- Scoop or tablespoon
- Water
- Instant *Drosophila* Medium (blue)
- Two vials of vestigial-winged (vg) culture of *Drosophila melanogaster*
- Several clean (preferably sterilized) pieces of filter paper
- Metric ruler
- *Drosophila* sorting brush or thin paintbrush
- Wax marking pencil
- UV-protective goggles
- UV light source
- Stopwatch or watch with a second hand

= Safety Icon

(continued)

Helpful Hints and Discussion

Time frame: Two weeks, including a single period of instruction the first day, and a few minutes of observation during each subsequent day. These observations may be made before or after class.
Structure: Groups of four students
Location: In class

If your class has not completed the preceeding activity ("From Egg to Adult: A Fruit Fly's Life "), please have them do so before beginning this activity. Students will be more successful in carrying out this activity if they are familiar with the life cycle of *Drosophila melanogaster* and have worked with this organism.

Although the scientific literature states that it takes two weeks for a *Drosophila melanogaster* egg to become an adult, the time may vary. Many conditions, such as temperature, amount of food, and the presence of other organisms, can influence both the length of the life cycle and the survival rate of the animal. In this activity, students investigate whether irradiation of *Drosophila melanogaster* larvae with ultraviolet light affects the length of the pupa stage and the survival rate. *Caution:* Warn your students not to look directly at the ultraviolet light and to wear protective goggles. Although these lights will be on for only one minute, prolonged exposure to ultraviolet rays can damage the eyes.

One of the problems that your students might encounter is contamination of the culture vials by mold. To avoid this, make sure that they work carefully and that all surfaces and instruments are disinfected with Lysol or a similar product before beginning.

This activity contains some mathematical concepts that are involved in data collection and calculations. Review these topics as part of your introduction.

Adaptations for High and Low Achievers

High Achievers: These students should help lower achievers by working with them in a group. Also, encourage these students to perform additional experiments based on follow-up activity 3.

Low Achievers: Provide a glossary and/or reference material. Heterogeneous groupings will enable students of higher ability to assist those having difficulty carrying out the activity.

Scoring Rubric

Full credit should be given to students who use complete sentences to correctly answer all the questions and accurately complete the three data tables provided. Extra credit should be given to students who complete any of the follow-up activities.

Internet Tie-ins

http://vflylab.calstatela.edu/
http://www-leland.stanford.edu/

Quiz

Base your answers to the following questions on the passage below and your knowledge of biology. Use complete sentences to answer the questions.

A student wants to discover how light affects the number of fruit fly larvae that will become pupae and then adults. She takes 10 larvae and puts them into a vial with prepared *Drosophila* food. She then covers the vial with a foam plug and leaves it under a fluorescent light for two weeks. At the end of two weeks, she notices that there are five pupae but no adults. She concludes that light is harmful to fruit flies and prevents them from growing to adulthood.

1. Do you agree with her conclusion? Why or why not?
2. How would you improve her experiment?
3. What factor besides light might have caused these results?

Name____ Date ____________________

UV Light: Effect on Hatching Time

STUDENT ACTIVITY PAGE

Although scientists tell us that it takes two weeks for a *Drosophila melanogaster* egg to become an adult, the time may vary. Many conditions, such as temperature, amount of food, or presence of other organisms, can influence both the length of the life cycle and the survival rate of the animal.

One factor that may have an effect is ultraviolet (UV) light. In recent years, we have become concerned that holes in the ozone layer are allowing more UV light to penetrate our atmosphere and harm living organisms. In this activity, you will investigate whether irradiation of *Drosophila melanogaster* larvae affects the length of the pupa stage and the survival rate. *Caution:* Although you will be exposing your organisms to ultraviolet light for only one minute, prolonged exposure to these rays can damage your eyes. Never look directly at the light and be sure to wear protective goggles.

You will be using flies with vestigial wings. Make sure that you work carefully and that all surfaces and instruments are clean. This will help prevent contamination by molds.

- Diluted disinfectant, such as Lysol, in a wash bottle
- Paper towels
- Clean probe
- Scoopula
- Three empty *Drosophila* culture vials, foam plugs, and vial caps
- Scoop or tablespoon
- Water
- Instant *Drosophila* Medium (blue)
- Two vials of vestigial-winged (vg) culture of *Drosophila melanogaster*
- Several clean (preferably sterilized) pieces of filter paper.
- Metric ruler
- *Drosophila* sorting brush or thin paintbrush
- Wax marking pencil
- UV-protective goggles
- UV light source
- Stopwatch or watch with a second hand

PROCEDURE

1. Wash your tabletop, probe, and scoopula with disinfectant. Dry the instruments and tabletop.
2. Prepare three vials of food in the following way: Into an empty vial, place one scoop or tablespoon of dry food. Add an equal amount of water. Swirl the contents gently to mix. When the mixture hardens (approximately one minute), add two or three grains of yeast. This will help prevent mold from forming. Immediately cover the vial with a foam plug.
3. Transfer all adult flies to one of the vials you just prepared.

(continued)

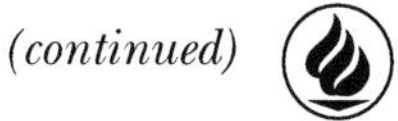

Name________________________________ Date ________________

UV Light: Effect on Hatching Time *(continued)*

STUDENT ACTIVITY PAGE

4. Using the scoopula, remove the food from the original vial and place it on the filter paper. Spread it out so that the larvae are visible. Select 20 of the smallest larvae. Return the rest of the larvae and food to the vial and replace the foam plug.
5. Divide the larvae into two groups of similar size.
6. Using your brush, place one group into a new vial with food. Label the vial with the date and the letter *A*. These are your control larvae.
7. Wearing your protective goggles, hold your UV light approximately 15 cm from the larvae remaining on the filter paper. Select a member of your cooperative learning group to be the timekeeper. Irradiate the larvae for one minute. Then place them into a new vial with food. Label the vial with the date and the letter *B*. These are the experimental larvae.
8. Observe the larvae daily for two weeks, keeping records of your observations. Record the dates of the appearance of pupae and of the emergence of adults in each vial in the Data Collection and Analysis section. Keep track of the time it takes for each pupa to become an adult. Also, look for differences in eye color, body color, and wing shape.

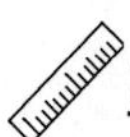

DATA COLLECTION AND ANALYSIS

1. Complete the following tables.

Table 1: Metamorphosis in Flies That Weren't Irradiated (A)

Control Group

Date	Day #	Larvae	Pupae	Adults
	1	10	0	0
	2			
	3			
	4			
	5			
	6			
	7			
	8			
	9			
	10			

(continued)

Name______________________________ Date ______________

UV Light: Effect on Hatching Time *(continued)*

STUDENT ACTIVITY PAGE

Table 2: Metamorphosis in Flies That Were Irradiated (B)

Experimental Group

Date	Day #	Larvae	Pupae	Adults
	1	10	0	0
	2			
	3			
	4			
	5			
	6			
	7			
	8			
	9			
	10			

2. Calculate the average length of the pupa stage in each vial. To do this, (a) determine the number of days it took for each pupa to become an adult, (b) add these numbers together, and (c) divide by the total number of adults.
3. Calculate the percentage of survivors in each vial by using the following formula:

   % Survivors = (Number of Adults/10) X 100

4. Complete the following summary table.

	Not Irradiated (A)	Irradiated (B)
Number of pupae		
Average length of pupa		
Stage		
Number of adults		
% survival		
Unusual adults?		
(Describe)		

5. Based on the tables and calculations above, draw as many conclusions as possible about the effect of ultraviolet light on the metamorphosis of *Drosophila melanogaster.* Write your answers in complete sentences.

(continued)

Name ______________________________ Date ______________

UV Light: Effect on Hatching Time *(continued)*

STUDENT ACTIVITY PAGE

Answer all questions on the back of this sheet or on a separate page.

1. Why was it necessary to separate the larvae into two similiar groups?
2. Why was it important that only one group be treated with ultraviolet light?
3. Why did you divide by 10 in the formula to determine the percentage of survivors?

Follow-up Activities

1. Present your data in the form of a bar graph.
2. Use your textbook, scientific journals, or the Internet to learn how ultraviolet radiation affects living cells. Report your results to the class.
3. Design an experiment to answer the following question: What is the effect of exposure time to ultraviolet light on metamorphosis in *Drosophila melanogaster*? Use times that range from 30 seconds to 5 minutes. After discussing your ideas with your teacher, perform the experiment.

© 1998 J. Weston Walch, Publisher

Complete Metamorphosis: A *Tenebrio's* Life Cycle

TEACHER RESOURCE PAGE

INSTRUCTIONAL OBJECTIVES

Students will be able to

- record observations of a living organism.
- list stages in the life cycle of an insect.
- draw and label stages of *Tenebrio*'s life cycle.
- formulate hypotheses.
- draw conclusions based on data.

NATIONAL SCIENCE STANDARDS ADDRESSED

Students demonstrate understanding of

- reproduction and heredity.
- big ideas and unifying concepts, such as form and function.

Students demonstrate scientific inquiry and problem-solving skills by

- identifying variables in experimental and nonexperimental research conditions.
- using evidence from reliable sources to develop descriptions and explanations.
- proposing, recognizing, and analyzing alternative explanations.
- working in teams to collect and share information and ideas.

Students demonstrate competence with the tools and technology of science by

- acquiring information from multiple sources, such as print, the Internet, and experimentation.

Students demonstrate effective scientific communication by

- representing data and results in multiple ways, such as drawings and technical writing.

MATERIALS

Each group of four students will need

- Tank containing *Tenebrio* in all stages of its life cycle, wheat bran, water supply
- Spoon or scoopula
- Petri dish
- Stereomicroscope
- Plastic gloves

HELPFUL HINTS AND DISCUSSION

Time frame: Single period of instruction
Structure: Groups of four students
Location: In class

Like many insects, the flour beetle, *Tenebrio molitar*, undergoes complete metamorphosis as it develops from egg to adult. In this activity, students observe these changes and draw some conclusions about the life cycle.

Tenebrio are inexpensive and very easy to maintain in the laboratory. They may be kept in a screen-covered tank that has a layer of wheat bran (approximately 2 cm) on the bottom. Provide water in a test tube covered with moist, absorbent cotton. A piece of apple or potato may be added occasionally. A culture that is several weeks old should have all four stages of the life cycle (egg, larva, pupa, adult) and should also have some shed exoskeletons produced by molting larvae.

When introducing the activity, tell your students that they will not be able to see eggs with the naked eye and may not see them even under the stereomicroscope. Tell them to look for eggs for only a few minutes and then go on to the next step. Also, stress that *Tenebrio* are living organisms and should be treated humanely. These slow-moving, harmless organisms may be picked up with the hands or a spoon or scoopula. Because some people may be allergic to insects, encourage your students to wear plastic gloves during the experiment and to wash their hands when they have finished.

(continued)

Adaptations for High and Low Achievers

High Achievers: These students should help lower achievers by working with them in a group. Also, encourage these students to perform additional behavioral experiments based on follow-up activities 1 and 2.

Low Achievers: Provide a glossary and/or reference material for the italicized terms in this activity. Heterogeneous groupings will enable students of higher ability to assist those having difficulty carrying out the activity. Encourage these students to perform additional experiments based on follow-up activity 3.

Scoring Rubric

Full credit should be given to students who use complete sentences to correctly answer all the questions and accurately complete the data table provided. Extra credit should be given to students who complete any of the follow-up activities.

www.birminghamzoo.com/ao/arthrop.htm (Animal Omnibus)
http://www.gene.com/ae/AE/AEC/AEF/index.html (Access Excellence)

Using the terms below, write a paragraph describing the changes that take place during each stage of a *Tenebrio*'s life cycle. In your description, mention the stages in order, starting with the earliest.

pupa   larva   adult   egg

Name____ Date __________________________

Complete Metamorphosis: A *Tenebrio*'s Life Cycle

STUDENT ACTIVITY PAGE

BEFORE YOU BEGIN

Like many insects, the flour beetle, *Tenebrio molitar,* undergoes a process called complete *metamorphosis* as it develops from egg to adult. Today you are going to observe these changes and draw some conclusions about its life cycle.

When handling *Tenebrio,* be gentle. They move rather slowly and may be picked up either with your hands or with a spoon or scoopula. Because some people may be allergic to certain insects, it is advisable to wear gloves if you touch any insect. Make sure that you wash your hands thoroughly after this laboratory activity.

MATERIALS

- Tank containing *Tenebrio* in all stages of its life cycle, wheat bran, water supply
- Spoon or scoopula
- Petri dish
- Stereomicroscope
- Plastic gloves

PROCEDURE

1. Observe the *Tenebrio* in their tank for five minutes. Although the organisms may look very different from each other, they are all *Tenebrio molitar,* in different stages of their life cycle. Record your observations. Draw as many different stages as you can identify. How many do you see? Label as many structures as possible.
2. The stage that you probably can't see is the *egg.* If you scoop some of the bran into the bottom of the petri dish and look at it under the stereomicroscope, you might be able to observe eggs. How would you expect their shape to be different from that of the flakes of bran? If you see an egg, draw it.
3. The egg will eventually develop into a *larva. Tenebrio*'s larva is called a mealworm. Pick one up and observe it carefully with both the naked eye and the stereomicroscope. What evidence do you have that it is an insect rather than a worm? Based on these observations, revise the drawing that you made in step 1.
4. You have probably seen larvae of different sizes in the tank. Look at the bottom of the tank for evidence that something has happened to each larva as it has grown. What evidence do you find? What conclusion(s) can you draw?
5. Look for an organism that is very still, or *sessile.* It may be lighter in color than the larva and be starting to develop a more rounded shape. This is the *pupa.* Place it in your petri dish and observe it under the stereomicroscope. What new structures do you see? If you did not draw a pupa in step 1, draw it now. If you did, revise your drawing based on your more careful observations.

(continued)

Name______________________________ Date ______________

Complete Metamorphosis: A *Tenebrio's* Life Cycle *(continued)*

STUDENT ACTIVITY PAGE

6. Look at both *dorsal* and *ventral* surfaces of the pupa. What evidence do you have that it will develop into an adult beetle?
7. Observe the *adult* flour beetles. Draw one. Label all structures that indicate that it is an insect.
8. When you have finished, return all organisms to their original tank. Clean up your station and wash your hands thoroughly.

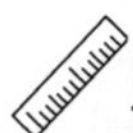

DATA COLLECTION AND ANALYSIS

1. Record your observations and first drawings of *Tenebrio*. Be as specific as possible.
2. In the table below, draw each of the stages of *Tenebrio's* life cycle that you saw. Label as many structures as possible.

Stage	Drawing
Egg	
Larva	
Pupa	
Adult	

(continued)

© 1998 J. Weston Walch, Publisher

Name________________________________ Date ____________________

Complete Metamorphosis: A *Tenebrio's* Life Cycle *(continued)*

STUDENT ACTIVITY PAGE

3. What evidence did you find in the tank that something happened to each larva as it grew? What conclusions can you draw from this evidence?
4. What evidence do you have that the pupa will develop into an adult beetle?

CONCLUDING QUESTIONS

Answer all questions on the back of this sheet or on a separate page.

1. Use your dictionary or glossary to define the word *metamorphosis.* Based on what you saw in this activity, why is *Tenebrio*'s life cycle called complete metamorphosis?
2. Insects that damage crops usually do the most damage during the larval stage. Why do you think this is true?
3. During which stage do you think insects do the least damage? Why?
4. A student finds an animal on the ground that looks like a worm. How could the student determine whether it is a worm or an insect larva?

FOLLOW-UP ACTIVITIES

1. Use a reference book or the Internet to learn about the life cycle of another insect that undergoes complete metamorphosis.
2. Design an experiment to determine the effect of light on the behavior of *Tenebrio* larvae.
3. Start with a colony of adult beetles. Take care of it and make observations for several weeks. Keep records of the changes and duration of each stage.

© 1998 J. Weston Walch, Publisher

Incomplete Metamorphosis: A Cricket's Life Cycle

TEACHER RESOURCE PAGE

INSTRUCTIONAL OBJECTIVES

Students will be able to

- record observations of living organisms.
- record data in a data table.
- draw conclusions based on data.
- formulate hypotheses about stages of a cricket's life cycle.
- draw and label the stages of a cricket's life cycle.

NATIONAL SCIENCE STANDARDS ADDRESSED

Students demonstrate understanding of

- structure and function in living systems.
- sexual reproduction.
- adaptation of organisms.
- big ideas and unifying concepts, such as form and function and cause and effect.

Students demonstrate scientific inquiry and problem-solving skills by

- framing both *cause* and *effect* questions.
- identifying problems and proposing and implementing solutions.
- working in teams to collect and share information and ideas.
- acquiring information from a variety of sources.
- evaluating the accuracy, design, and outcomes of investigations.
- using technology and tools to observe organisms and phenomena.

Students demonstrate effective scientific communication by

- using evidence from reliable sources to develop descriptions, explanations, and models.
- recording data in a variety of formats.

MATERIALS

Each group of four students will need

- Tank containing male and female crickets, cricket nutrient, water supply, piece of lettuce, apple, or potato
- Petri dish
- Moist sand
- Stereomicroscope
- Plastic gloves

HELPFUL HINTS AND DISCUSSION

Time frame: One month
Structure: Individuals or cooperative learning groups
Location: In class or at home

In this activity, students make daily observations of developing eggs of the house cricket, *Acheta domestica,* and record changes. This activity will take approximately one month and except for the first observations will require only a few minutes each day. Encourage students to get involved in another activity at the same time. There may be some days when nothing seems to be happening. Then all of a sudden, exciting changes will occur. It's worth the wait!

To make accurate observations, it will be necessary to use a stereomicroscope.

Take a few minutes during your introduction to demonstrate how to use this instrument.

(continued)

Adaptations For High And Low Achievers

High Achievers: These students should help lower achievers by working with them in a group. Also, encourage these students to perform additional experiments based on concluding question 5.

Low Achievers: Provide a glossary and/or reference material for the italicized terms in this activity. Encourage these students to complete follow-up activity 3. Heterogeneous groupings will enable students of higher ability to assist those having difficulty carrying out the activity.

Scoring Rubric

Full credit should be given to students who comple the journal and data table and correctly answer all the questions in complete sentences. Extra credit should be given to students who complete any of the follow-up activities.

Internet Tie-ins

http://www.birminghamzoo.com/ao/arthrop.htm (Animal Omnibus)
http://www.colostate.edu/Depts/Entomology/ent.htm
http://viceroy.eeb.uconn.edu/OS_Homepage (Orthopterists' Society)
http://www.gene.com/ae/AE/AEC/AEF/index.html (Access Excellence)

Quiz

1. How is the cricket nymph similar to the adult? How is it different?
2. Why is it necessary for the cricket nymph to molt (shed its exoskeleton) several times before it becomes an adult?

Name________________________________ Date ____________________

Incomplete Metamorphosis: A Cricket's Life Cycle

STUDENT ACTIVITY PAGE

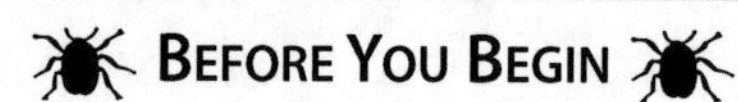

During the next few weeks, you will observe the life cycle of the house cricket, *Acheta domestica.* Look for changes that occur as the crickets develop from eggs to adults. You will also make observations about the length and stages of their life span. This activity takes only a few minutes each day but will take approximately one month to complete. As you can imagine, it will require great patience. There may be some days when nothing seems to be happening, but do not be discouraged. All of a sudden, you will start to see exciting changes. Be prepared.

- Tank containing male and female crickets, cricket nutrient, water supply, piece of lettuce, apple, or potato
- Petri dish
- Moist sand
- Stereomicroscope
- Plastic gloves

PROCEDURE

1. Observe the crickets in their tank. How many males are there? Females? How can you tell them apart? Draw pictures of male and female crickets.
2. Fill the bottom half of the petri dish with moist sand. Look at the sand carefully. Draw a picture of the dish of sand. What color(s) do you see?
3. Now place the dish of sand on the stage of a stereomicroscope. Adjust the focus knob so that you can clearly see each grain. Draw a picture of the magnified grains of sand. What is the shape of the grains? What color(s) are they? Are the edges straight or rounded?
4. Place the dish of moist sand in the tank. Observe the crickets. During the course of this activity, try to observe the crickets as often as possible. In particular, look at their behavior in and around the dish of sand. If you see a female laying eggs, take careful notes and be ready to describe this to your class. When there are eggs in the dish, how will you be able to distinguish them from the grains of sand? What shape do you expect? Size? Color?
5. Several times a week, remove the dish of sand from the tank. Cover it with the top half of the petri dish and observe it under the stereomicroscope. Draw what you see and describe any changes, using one or two complete sentences. Put the dish back in the tank and remove the top of the petri dish.
6. In two to three weeks, you might see young crickets, called *nymphs,* who can move. As soon as you see this, transfer the dish to a fresh tank. Explain why this is necessary. Continue your observations. Draw pictures of the nymphs. How do they compare with the adults?
7. Like all insects, crickets have a nonliving *exoskeleton* made of a chemical called *chitin.* In order to grow, they must shed their exoskeleton. Look for evidence of exoskeletons that have been shed. Using plastic gloves, carefully pick up a shed exoskeleton and examine it under the stereomicroscope. Which parts of the insect are covered by the exoskeleton?
8. After each day's observations, clean up your lab station and thoroughly wash your hands.

(continued)

Name____ Date ______________________

Incomplete Metamorphosis: A Cricket's Life Cycle *(continued)*

STUDENT ACTIVITY PAGE

DATA COLLECTION AND ANALYSIS

1. Record your daily observations for three weeks in a laboratory notebook or journal. For each entry, indicate the following: *Date, Drawing* or *Diagram, Description.* Make sure to write your descriptions in one or more complete sentences.
2. Based on your observations and journal, complete the following table.

Day #	Eggs in Sand (Y/N)	Nymphs in Sand (Y/N)	Nymphs in Tank (Y/N)

CONCLUDING QUESTIONS

Answer all questions on the back of this sheet or on a separate page.

1. Refer back to your prediction about the size, shape, and color of the crickets' eggs. Explain if your hypothesis was correct.
2. How much care did the adult crickets give to the developing eggs? What effect do you think this will have on the number of eggs that will survive?
3. What was your first clue that an egg was developing into an insect?
4. How did the cricket nymph compare with the adult cricket in size, shape, and color?
5. Use your textbook or another reference source to learn about the life cycle of a butterfly. Why is the life cycle of the butterfly *complete metamorphosis* while the life cycle of the cricket is *incomplete metamorphosis*?

FOLLOW-UP ACTIVITIES

1. At what point in the life cycle would you say that a nymph has become an adult?
2. Draw a time line showing the stages of a cricket's development from egg to adult.
3. Make an artistic display illustrating the life cycle of crickets. You might want to use shed exoskeletons as part of your display.

© 1998 J. Weston Walch, Publisher

Observing Insects: How Do Crickets Behave?

Teacher Resource Page

Instructional Objectives

Students will be able to

- record observations of a living organism.
- list life functions that are observed.
- record data in a data table.
- draw conclusions based on data
- formulate hypotheses about aspects of cricket behavior.
- draw and label a cricket.

National Science Standards Addressed

Students demonstrate understanding of

- structure and function in living systems.
- regulation of behavior.
- adaptation of organisms.
- big ideas and unifying concepts, such as form and function and cause and effect.

Students demonstrate scientific inquiry and problem-solving skills by

- framing questions.
- idendifying varables.
- identifying problems; proposing and implementing solutions.
- Working in teams to collect and share information and ideas.
- Recognizing sources of bias in data.

Students demonstrate effective scientific communication by

- Recording data in a variety of formats.

Materials

Each group of four students will need

- Tank containing male and female crickets, cricket nutrient, dish of moist sand, and water supply
- Plastic gloves
- Plastic vial with foam plug
- Empty tank
- Probe or pencil
- Piece of lettuce, apple, or potato
- Shallow dish containing a thin layer of flour

Helpful Hints and Discussion

In this activity, students will assume the role of behavioral scientists by observing and reporting on cricket behavior. Crickets are inexpensive and easy to maintain in the laboratory. They may be purchased from biological supply houses or local pet shops. They should be maintained at room temperature in a dry tank, which may be covered with screening. To prevent crickets from escaping you might want to put a 3 cm-thick ring of petroleum jelly along the rim of the tank before adding the screen. The bottom of the container should be covered with 2- to 4-cm-thick layer of wood shavings, sawdust, or sand. The crickets will also need a small dish of food (cricket nutrient or chopped dog food) as well as pieces of lettuce, apple, or potato. They also need a water supply. A small test tube filled with water and covered with moist, absorbent cotton works very well for this purpose. Finally, provide a dish of moist sand in which the crickets can lay eggs (the top or bottom of a petri dish is perfect). To complete the crickets' home, add a small peat flowerpot or section of a cardboard egg carton as additional living space. Keep crickets clean, change the water often, and remove waste and uneaten food weekly.

When introducing this activity, emphasize the following information about the techniques for observing organisms.

- The behavior of organizms is often unpredictable. Therefore, observing, collecting results, and drawing conclusions about organisms in their habitat require patience and an open mind.
- Sudden movements or loud noises will interfere with experiments.
- Organisms are alive and should be treated humanely.
- Wear plastic gloves when hangling organisms to avoid allergic reactions. Wash hands thoroughly after the experiment.

(continued)

Adaptations for High and Low Achievers

High Achievers: These students should help lower achievers by working with them in a group. Also, encourage these students to perform additional behavioral experiments based on concluding question 3.

Low Achievers: Provide a glossary and/or reference material for the italicized terms in this activity. Heterogeneous groupings will enable students of higher ability to assist those having difficulty carrying out the activity.

Scoring Rubric

Full credit should be given to students who use complete sentences to correctly answer all the questions and accurately complete the data tables provided. Extra credit should be given to students who complete any of the follow-up activities.

Internet Tie-ins

http://www.birminghamzoo.com/ao/arthrop.htm (Animal Omnibus)
http://www.colostate.edu/Depts/Entomology/ent.htm
http://viceroy.eeb.uconn.edu/OS_Homepage (Orthopterists' Society)
http://www.gene.com/ae/AE/AEC/AEF/index.html (Access Excellence)

Quiz

1. Using the information you have collected during this activity, write a short essay describing how crickets perform each of the following life activities.

   (a) nutrition
   (b) regulation
   (c) locomotion
   (d) reproduction

2. What evidence do you have that the cricket is an insect?

Name____ Date ____________________

Observing Insects: How Do Crickets Behave?

STUDENT ACTIVITY PAGE

BEFORE YOU BEGIN

Today you will investigate the behavior of the house cricket, *Acheta domestica.* This insect, like its relative the grasshopper, can jump and sing. It is herbivorous, with mouthparts that can chew plant material. It is very easy to distinguish between male and female crickets because the females have a long tube, called an *ovipositor,* on their posterior segment (see Figures 1a and 1b, drawings of male and female crickets).

When studying animal behavior, there are a few rules to follow. When moving the animal, be careful not to harm it and give it time to recover. Be patient and do not make any sudden noises or movements once your observations have begun. Remember that the animal may not be hungry when you choose to feed it or may not behave as you expect. If this happens, try your experiment again at another time. Above all else, keep an open mind and enjoy your observations of animal behavior.

The crickets that you will observe have been kept in a tank at room temperature. They have been provided with water, food, and a dish of moist sand in which to lay eggs. Their tank also contains a small peat flowerpot or a piece of egg carton as additional living space.

Be gentle when handling the crickets. An easy way to move them from place to place is to use a small plastic vial or jar fitted with a foam cover. Place the open vial on its side in the tank and try to capture a cricket. Immediately cover the vial with the foam. After a few tries you will become an expert. Remember, these insects are harmless and will not bite humans. Because some people may be allergic to certain insects, it is advisable to wear gloves if you touch any insect. Make sure that you wash your hands thoroughly after this laboratory activity.

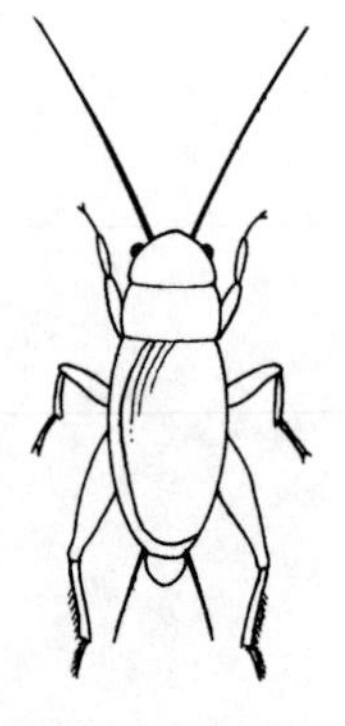

Figure 1a
male cricket

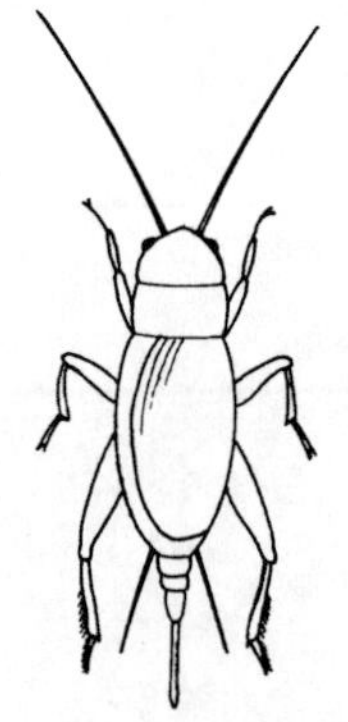

Figure 1b
female cricket

(continued)

Name____ Date ________________________

Observing Insects: How Do Crickets Behave? *(continued)*

STUDENT ACTIVITY PAGE

MATERIALS

- Tank containing male and female crickets, cricket nutrient, dish of moist sand, and water supply
- Plastic gloves
- Plastic vial with foam plug
- Empty tank
- Probe or pencil
- Piece of lettuce, apple, or potato
- Shallow dish containing a thin layer of flour

PROCEDURE

1. Observe the crickets in their tanks for five minutes. Take notes on your observations. How many males and how many females are in your tank? How do the crickets move? Do they appear to be communicating? How can you tell? How do you know that they are insects?
2. Capture one male and put it in the empty tank. What sense organs does it use to become familiar with its new environment? After a few minutes, gently tap a probe or pencil approximately one inch in front of it, to the right, to the left, and behind it. Describe how it reacts to these stimuli. Explain why you think the cricket responded as it did.
3. Add a piece of lettuce, apple, or potato to the tank. What does the cricket do? If the cricket eats, observe the way it ingests the food. Record your observations. Return the male to its original tank.
4. Repeat step 2, using a female instead of a male. Describe any differences you observe. Return the female to its original tank.
5. Put another male in the new tank. After a few minutes, add a second male. Observe them carefully and record what happens when they sense each other. Does one seem dominant? Explain.
6. Now add a female to the new tank. Observe the behaviors of all the crickets carefully and explain what happens. Use the terms *dominant male* and *submissive male* in your explanation. Return the crickets to their original tank.
7. Put a shallow dish of flour into the empty tank. Put a new cricket in the dish, allowing its legs and antennae to touch the flour. Then take the cricket out of the flour and place it on the floor of the tank. What does the cricket do? Explain why this is called *grooming behavior.*
8. After completing your observations, return all crickets to their original tank. Clean your lab station and thoroughly wash your hands. If time permits, discuss your results with the other members of your group.

(continued)

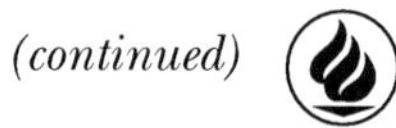

© 1998 J. Weston Walch, Publisher

Name________________________________ Date ____________________

Observing Insects: How Do Crickets Behave? *(continued)*

STUDENT ACTIVITY PAGE

DATA COLLECTION AND ANALYSIS

1. Record your observations of the behavior of crickets in their original tank. Complete the following table.

Tap Location	Male's Reaction	Female's Reaction
Front		
Back		
Left Side		
Right Side		

2. On the diagram below, place an *X* where you think the cricket has a sense organ to detect motion.

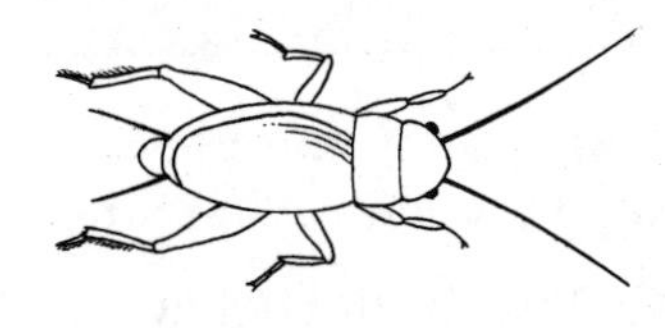

3. Describe how the cricket ingests food. Draw a picture to illustrate this process. Label the parts that are involved in ingestion.

4. Compare the female's reaction to stimuli to the male's.

5. Describe the interactions between the two males.

What Male 1 Did	What Male 2 Did

(continued)

Name______________________________ Date ______________________

Observing Insects: How Do Crickets Behave? *(continued)*

Student Activity Page

6. Describe the interactions of the crickets when the female was added to the tank with the two males.

What Male 1 Did	What Male 2 Did	What Female Did

Explain the chart you have just made, using the terms *dominant male* and *submissive male.*

7. Describe the grooming behavior of the cricket.

Concluding Questions

Answer all questions on the back of this sheet or on a separate page.

1. Why would a behavioral scientist want to repeat steps 5 and 6 several times before drawing any conclusions?
2. Why might the experiment work better if the crickets are isolated from each other for at least one day?
3. What are two questions about cricket behavior that have not been answered by this activity?

Follow-up Activities

1. Repeat steps 2 and 4, measuring how high and far the crickets jump.
2. If you had three male crickets, how would you test to find out which was the most dominant? Least dominant?
3. Design an experiment to answer one of your unanswered questions from concluding question 3.

© 1998 J. Weston Walch, Publisher

Adaptations: How Insects Escape

TEACHER RESOURCE PAGE

INSTRUCTIONAL OBJECTIVES

Students will be able to

- record observations of living organisms.
- compare and contrast behavior of two different species of insects.
- measure distance using the metric system.
- record data in a data table.
- perform simple calculations.
- interpret data and draw conclusions.

NATIONAL SCIENCE STANDARDS ADDRESSED

Students demonstrate understanding of

- structure and function in living systems.
- regulation and behavior.
- adaptation of organisms.
- big ideas and unifying concepts, such as form and function and cause and effect.

Students demonstrate scientific inquiry and problem-solving skills by

- identifying variables in experimental research settings.
- evaluating the outcomes of investigations.
- working in teams to share information and ideas.

Students demonstrate competence with the tools and technologies of science by

- using technology and tools to observe and measure organisms directly and indirectly.
- recording and storing data.
- collecting and analyzing data.

Students demonstrate effective scientific communication by

- arguing from evidence, such as data produced through experimentation.

MATERIALS

Each group of four students will need

- Tank of male and female crickets
- Track made of wood, cardboard, or plastic (55 cm long X 4 cm wide X 20 cm high)
- Piece of cardboard 4 cm wide X 5 cm high to serve as "gate" at 5-cm mark on track
- Paper grid with lines 1 cm apart to line the bottom and sides of the track
- Plastic vial and foam plug
- Plastic gloves
- Stopwatch or watch with a second hand
- Plastic medicine dropper
- Tank of male and female hissing cockroaches
- Paper towels

(continued)

Helpful Hints and Discussion

Time frame: Single period of instruction
Structure: Groups of four students
Location: In class

In this activity, students perform experiments to discover which organs in the house cricket, *Acheta domesticus,* and the Madagascan hissing cockroach, *Gromphodorina portentosa,* detect pressure. If resources are limited, you may choose to use only one of the organisms, or you may experiment with other insects that are available.

Both the house cricket and the Madagascan hissing cockroach are easy to maintain and require similar conditions. They should be kept in screen-covered tanks at room temperature with a supply of water and food (dry dog food for cockroaches and a commercial nutrient preparation for crickets). A small peat flowerpot or piece of egg carton for additional living space should be placed in each tank. An occasional slice of apple or potato may be provided. The crickets will also need a dish of moist sand in which to lay eggs.

Before your students begin this activity, demonstrate methods of transferring the animals. An easy way to move the crickets from place to place is to use a small plastic vial or jar fitted with a foam cover. Place the open vial on its side in the tank and try to capture a cricket. Immediately cover the vial with the foam. An easy way to handle the cockroaches is to pick them up and place them in your hand. After a few tries, your students will be able to do this easily. Emphasize that these animals are alive and should be treated humanely. The cockroaches move *very* slowly and are completely harmless to humans. Because some people may be allergic to certain insects, it is advisable for your students to wear gloves if they touch any insect. Make sure that they wash their hands thoroughly after this laboratory activity.

Adaptations for High and Low Achievers

High Achievers: These students should help lower achievers by working with them in groups. Also, encourage these students to perform additional behavioral experiments based on the follow-up activities.

Low Achievers: Provide a glossary and/or reference material for the italicized terms in this activity. Review the relevant anatomy of the insects before students begin the activity. Heterogeneous groupings will enable students of higher ability to assist those having difficulty carrying out the activity. Students who are low achievers in science but higher achievers in mathematics might perform calculations for the group.

Scoring Rubric

Full credit should be given to students who use complete sentences to correctly answer all the questions and accurately complete the two data tables provided. Extra credit should be given to students who complete any of the follow-up activities.

www.carolina.com/tips/mar95/index.html (Carolina Biological Supply Co.)
http://viceroy.eeb.uconn.edu/OS_Homepage (Orthopterists' Society)
http://www.gene.com/ae/AE/AEC/AEF/index.html (Access Excellence)

Base your answers to the following questions on the paragraph below and your knowledge.

Student A wants to find out whether the cerci of cockroaches detect differences in air pressure. She directs a jet of air at the cerci and finds that the cockroach moves 50 cm in five seconds. She concluded that the cerci did detect the air pressure. Student B says that Student A needs to perform one more test in order to reach that conclusion.

1. What test would Student B suggest to Student A?
2. Why would such a test be necessary?

Name________________________ Date ____________

Adaptations: How Insects Escape

STUDENT ACTIVITY PAGE

BEFORE YOU BEGIN

If you have ever tried to capture an insect, you probably noticed that it seems to sense your presence. As you get closer it moves faster to get away. Insects have two sense organs that receive *tactile* (touch) stimuli—*antennae* and *cerci* (see Figure 1). The antennae are on the head, and the cerci are at the tip of the abdomen. Both organs contain sensitive hairs on an *appendage.*

In this activity, you will try to determine which of these organs detects differences in air pressure, and how the insect reacts to these differences. You will measure motion in the house cricket, *Acheta domesticus,*and the Madagascan hissing cockroach, *Gromphodorina portentosa.* Both species are easy to maintain and require similar conditions. They should be kept in screen-covered tanks at room temperature with a supply of water and food (dry dog food for cockroaches and a commercial nutrient preparation for crickets). A small peat flowerpot or piece of egg carton for additional living space should be placed in the tank. An occasional slice of apple or potato may be provided. The crickets will also need a dish of moist sand in which to lay eggs.

Figures 1 and 2 show you how to distinguish between males and females of both species. An easy way to move the crickets from place to place is to use a small plastic vial or jar fitted with a foam cover. Place the open vial on its side in the tank and try to capture a cricket. Immediately cover the vial with the foam. After a few tries you will become an expert.

An easy way to handle the cockroaches is to pick them up and place them in your hand. After a few tries, you will be able to do this easily. They move *very* slowly and are completely harmless to humans. Because some people may be allergic to certain insects, it is advisable to wear gloves if you touch any insect. Make sure that you wash your hands thoroughly after this laboratory activity.

Remember that these animals are alive and should be treated humanely. When moving the animals, be careful not to harm them and give them time to recover. Do not make any sudden noises or movements.

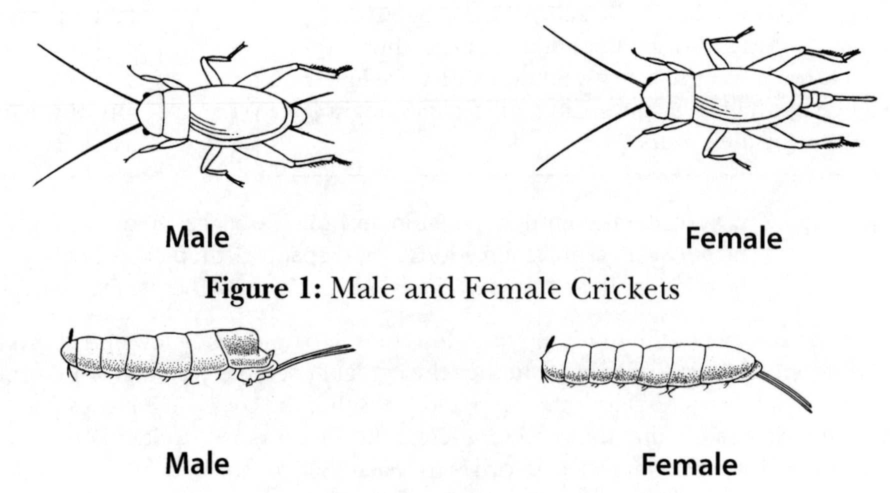

Figure 1: Male and Female Crickets

Figure 2: Male and Female Madagascan Hissing Cockroaches[1]

[1]http://www.carolina.com/tips/mar95/index.html (*Carolina Tips,* March 1995)

(continued)

Name____ Date ________________________

Adaptations: How Insects Escape *(continued)*

STUDENT ACTIVITY PAGE

MATERIALS

Each group of four students will need

- Tank of male and female crickets
- Track made of wood, cardboard, or plastic (55 cm long X 4 cm wide X 20 cm high)
- Piece of cardboard 4 cm wide X 5 cm high to serve as "gate" at 5-cm mark on track
- Paper grid with lines 1 cm apart to line the bottom and sides of the track
- Plastic vial and foam plug
- Plastic gloves
- Stopwatch or watch with a second hand
- Plastic medicine dropper
- Tank of male and female hissing cockroaches
- Paper towels

PROCEDURE

1. Assign the members of your team: animal handler, gatekeeper, timekeeper, and recorder.
2. Capture a male cricket and put it behind the "gate" in the track. Lift the gate and record the time it takes for the cricket to reach the other end of the track. If the cricket does not reach the end of the track, record the distance it traveled. If the cricket jumps, record the height of the jump by placing a pencil mark in the appropriate spot on the grid on the side of the track. Record your results in the Data Collection and Analysis section (Trial 1).
3. After the cricket has rested briefly, place it behind the gate. Hold your medicine dropper near the outside of the gate. Lift the gate. Gently squeeze the bulb of the medicine dropper so that a jet of air hits the cricket's antennae. Record the time it takes for the cricket to reach the other end of the track and the height of any jumps. Record your results in Table 1 (Trial 3).
4. Repeat step 3, but this time aim the jet of air at the cerci. Record your results in the Data Collection and Analysis section (Trial 5).
5. Repeat steps 2, 3, and 4 with a second male cricket. Record your results in the Data Collection and Analysis section (Trials 2, 4, 6).
6. Return both crickets to their tank.
7. Repeat steps 1 through 6, using male cockroaches instead of crickets.
8. When you have completed this activity, clean up and thoroughly wash your hands.

(continued)

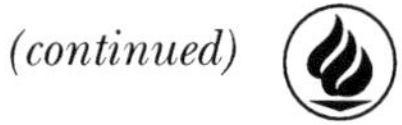

Name ______________________ Date ______________________

Adaptations: How Insects Escape *(continued)*

STUDENT ACTIVITY PAGE

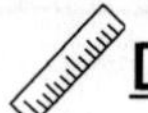

DATA COLLECTION AND ANALYSIS

1. Complete the table below.

Table 1: Motion of Male Crickets in Response to Currents of Air

Trial #	Time (seconds)	Number of jumps	Height of highest jump (cm)
1 (No air current — Cricket 1)			
2 (No air current — Cricket 2)			
3 (Air on antennae — Cricket 1)			
4 (Air on antennae — Cricket 2)			
5 (Air on cerci — Cricket 1)			
6 (Air on cerci — Cricket 2)			

2. Based on your data, explain which organs detect changes in air pressure.
3. Under which of the above conditions are crickets more likely to jump? What evidence do you have to support your answer?
4. Complete the table below.

Table 2: Motion of Male Cockroaches in Response to Currents of Air

Trial #	Time (seconds)
1 (No air current — Cockroach 1)	
2 (No air current — Cockroach 2)	
3 (Air on antennae — Cockroach 1)	
4 (Air on antennae — Cockroach 2)	
5 (Air on cerci — Cockroach 1)	
6 (Air on cerci — Cockroach 2)	

(continued)

Name________________________________ Date ____________________

Adaptations: How Insects Escape *(continued)*

Student Activity Page

5. Based on your data, explain where the sense organs that detect changes in air pressure are located in cockroaches.
6. Compare your results with those obtained by other groups in your class. How do you explain any differences?

Concluding Questions

Answer all questions on the back of this sheet or on a separate page.

1. Why was the procedure repeated with a second animal of the same species?
2. Why were both animals of the same sex?
3. How might the ability to detect changes in air pressure help an insect survive?
4. Why are Trials 1 and 2 called the controls?
5. If an insect travels 50 cm in 23 seconds, what is its speed in cm/sec?

Follow-up Activities

1. Calculate the average speed (cm/sec) for crickets and cockroaches under each of the three conditions that were tested. Design a data table that shows your information in a logical manner.
2. Design an experiment to determine whether other stimuli, such as food and light, affect the speed of a cricket or cockroach.
3. Use your textbooks, other printed materials, or the Internet to learn more about sense organs in insects. Report your findings to the class.
4. Repeat this activity using females instead of males. Are your results the same?
5. Hypothesize what would happen if you used animals of both sexes in this activity. Test your hypothesis.

© 1998 J. Weston Walch, Publisher

Measuring Movement: How High Can a Fruit Fly Climb?

TEACHER RESOURCE PAGE

INSTRUCTIONAL OBJECTIVES

Students will be able to

- record observations of a living organism.
- measure distance using the metric system.
- record data in a data table.
- perform simple calculations.
- interpret data and draw conclusions.

NATIONAL SCIENCE STANDARDS ADDRESSED

Students demonstrate understanding of

- regulation and behavior.
- adaptation of organisms.
- big ideas and unifying concepts, such as form and function and cause and effect.

Students demonstrate scientific inquiry and problem-solving skills by

- identifying or controlling variables in experimental research settings.
- evaluating the accuracy, design, and outcomes of investigations.
- working individually and in teams to collect and share information and ideas.

Sudents demonstrate competence with the tools and technologies of science by

- using technology and tools to observe and measure organisms directly and indirectly.
- recording and storing data.
- collecting and analyzing data using concepts of Mathematics Standards, such as mean.
- recognizing sources of bias in data, such as observer and sampling biases.

Students demonstrate effective scientific communication by

- arguing from evidence, such as data produced through experimentation.

MATERIALS

Each group of four students will need

- Vial of vestigial-winged (vg) culture of *Drosophila melanogaster*
- Two empty vials with foam plugs
- *Drosophila* sorting brush or thin paintbrush
- 100-ml graduated cylinder, preferably made of clear plastic
- Transparent glass or plastic square to cover graduated cylinder
- Metric ruler
- Stopwatch or watch with a second hand
- Light source

HELPFUL HINTS AND DISCUSSION

Time frame: Single period of instruction
Structure: Groups of four students
Location: In class

In this activity, students observe and measure the ability of the fruit fly, *Drosophila melanogaster*, to climb. They also determine whether this ability is affected by light.

Drosophila are inexpensive and easy to maintain in the laboratory. They may be purchased from many biological supply companies, along with materials for their maintenance. To prepare a culture, place approximately one tablespoon of Instant Drosophila Medium and one tablespoon of water in a culture vial. When the medium solidifies (approximately one minute), add a few grains of yeast. If you are not performing genetics or population experiments, there is no need to anesthetize the flies before transferring them. Simply follow directions in step 1 of the procedure on the student activity page. Several males and females should be moved into the new vial. Cover both vials with foam plugs. The flies can survive without additional care for several weeks.

Flies with vestigial wings are used in this experiment because they cannot fly and are easier to handle. If they do get out of the vial during a transfer, they can be swept back with a small paintbrush. Please remind your students that these organisms are small and fragile and should be handled carefully. Before allowing the students to proceed, demonstrate the method of transfer and have them practice the manipulation with two empty vials.

The concept of converting *volume* lines on the graduated cylinder to units of length might confuse your students. Take time to review this carefully.

(continued)

Adaptations for High and Low Achievers

High Achievers: These students should help lower achievers by working with them in a group. Also, encourage these students to perform additional behavioral experiments based on any of the follow-up activities.

Low Achievers: Provide a glossary and/or reference material for the italicized terms in this activity. Heterogeneous groupings will enable students of higher ability to assist those having difficulty carrying out the activity. Students who are low achievers in science but higher achievers in mathematics might perform calculations for the group.

Scoring Rubric

Full credit should be given to students who use complete sentences to correctly answer all the questions and accurately complete the two data tables provided. Extra credit should be given to students who complete any of the follow-up activities.

http://www.gene.com/ae/AE/AEC/AEF/index.html (Access Excellence)
http://www.carolonia.com (Carolina Biological Supply Co.)
http://vflylab.calstatela.edu/edesktop/VirtApps/VflyLab/IntroVflyLab.html
http://www-leland.stanford.edu/~ger/drosophila.html

Base your answers to the following questions on the information below. Answer all questions in complete sentences.

A student wants to determine whether temperature affects a fruit fly's ability to climb. He puts one vial of adult flies with vestigial wings in the refrigerator and leaves a vial of flies with long wings at room temperature. He then puts four flies from each vial into a large graduated cylinder and measures how high each one climbs. The average height climbed by the flies left in the refrigerator is 10 cm. The average height climbed by the flies at room temperature is 15 cm. Therefore, he concludes that flies can climb higher at higher temperatures.

1. Is the student's conclusion accurate? Why or why not?
2. What changes would you make in the design of the experiment?

Name________________________________ Date ________________

Measuring Movement: How High Can a Fruit Fly Climb?

STUDENT ACTIVITY PAGE

BEFORE YOU BEGIN

The fruit fly, *Drosophila melanogaster*, can move in several ways. Most *Drosophila* can fly; all can walk and climb. Today you will observe and measure *Drosophila*'s ability to climb and determine whether light affects this ability. You will be using flies with *vestigial* wings. Flies with these reduced wings cannot fly and are easier to handle.

When working with fruit flies, there are a few precautions to take. They are small and fragile, so be careful with them. When transferring them from one bottle (vial) to another, make sure that there is no space between the vials. This will prevent the flies from escaping. If they do get out of the vial, simply sweep them back with a small paintbrush. They will not harm you and move slowly enough that you can easily capture them.

The flies that you will observe have been kept in a vial that is covered with a sponge plug. This allows air to enter and leave. The food at the bottom of the vial contains equal amounts of Instant *Drosophila* Medium and water. A few grains of yeast have been added to prevent mold from growing. The flies can survive in this container for several weeks.

MATERIALS

- Vial of vestigial-winged (vg) culture of *Drosophila melanogaster*
- Two empty vials with foam plugs
- *Drosophila* sorting brush or thin paintbrush
- 100-ml graduated cylinder, preferably made of clear plastic
- Transparent glass or plastic square to cover graduated cylinder
- Metric ruler
- Stopwatch or watch with a second hand
- Light source

PROCEDURE

1. Read all directions in step 1 before beginning. Practice several times with two empty vials before trying to transfer flies.

   Transfer some vestigial flies to an empty vial by the following method:

   (a) Remove the foam plug from the empty vial and turn the vial upside down.

   (b) Take the vial that contains the flies and gently tap the bottom against the tabletop. This should cause the flies to fall to the surface of the food.

   (c) Place the empty vial in one hand. *Very quickly* remove the foam plug from the vial that contains the flies and place the empty inverted vial on top of it.

   (d) After at least four flies have moved into the empty vial, quickly replace both foam plugs.

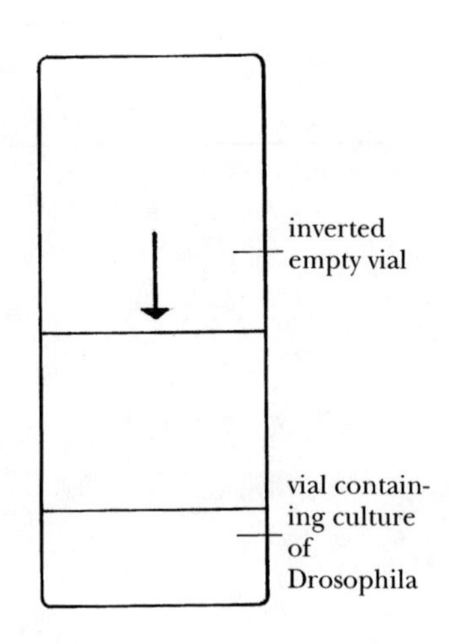

(continued)

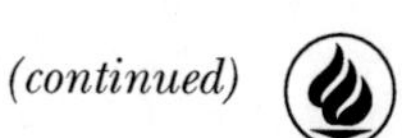

Name____ Date ______________________

Measuring Movement: How High Can a Fruit Fly Climb? *(continued)*

STUDENT ACTIVITY PAGE

2. Transfer the flies that moved into the empty vial into the 100-ml graduated cylinder by "pouring" them into the cylinder. Cover the cylinder with a glass or plastic square. Cover the empty vial with a foam plug.
3. Observe the flies for two minutes. Take notes on your observations. Describe the flies' motion. Try to determine the order in which they move their legs.
4. Very gently, tap the cylinder to cause the flies to fall to the bottom. Each member of your group will choose one fly, observe it for 10 seconds, and record the highest line that the fly reached during that time (Trial 1).

   Note: The graduated cylinder is usually used to measure *volume,* not length. Volume is measured in *milliliters,* while length is measured in *centimeters.*

   Therefore, it is necessary to determine how many centimeters the fly climbed. To do this, use your centimeter ruler to measure the height the fly reached. Enter your data in centimeters in the table in the Data Collection and Analysis section.
5. Repeat step 4, shining a light source at the top of the cylinder (Trial 2).
6. Using a different set of flies, repeat the experiment. This time, however, shine the light the first time (Trial 3) instead of the second (Trial 4).
7. When you have finished, clean up your laboratory space and carefully return all flies to their original vial.

DATA COLLECTION AND ANALYSIS

1. Record your observations of the behavior of the flies after they were transferred to the graduated cylinder. Be as specific as possible. Before submitting your observations to your teacher, revise your writing. Make sure that you have answered in complete sentences and checked your spelling.
2. Complete the following table.

Distance Flies Climbed Without Added Light Source

Trial #	Fly #	Line Reached in 10 Sec. (ml)	Actual Distance (cm)
1			
4			

3. Calculate the average distance, in centimeters, covered by the flies without added light. Use the following formula:

$$\frac{\textbf{Sum of distances (cm)}}{\textbf{Number of flies}} = \textbf{Average distance (cm)}$$

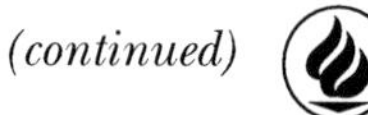
(continued)

© 1998 J. Weston Walch, Publisher

Name____ Date ____________________________

Measuring Movement: How High Can a Fruit Fly Climb? *(continued)*

STUDENT ACTIVITY PAGE

4. Complete the following table.

Distance Flies Climbed with Added Light Source

Trial #	Fly #	Line Reached in 10 Sec. (ml)	Actual Distance (cm)
2			
3			

5. Using the formula in step 3, calculate the average distance covered by flies when a light source was added.
6. Based on the data you collected, explain whether light affects the distance to which *Drosophila melanogaster* can climb.

CONCLUDING QUESTIONS

Answer all questions on the back of this sheet or on a separate page.

1. Why did your group use more than one fly in each trial?
2. Why was step 6 of the procedure necessary?
3. Calculate the average speed at which the flies climbed both with and without added light (average speed = average distance divided by time). Include the proper units in your answer.

FOLLOW-UP ACTIVITIES

1. Design an experiment to determine whether male or female flies can climb higher. What additional information do you need in order to carry out this experiment?
2. Calculate the average speed of the flies in kilometers per hour.
3. Design an experiment to answer the following question: What is the effect of ________________________ on the ability of *Drosophila melanogaster* to climb?
4. Both insects and humans use muscles to move. However, there are differences in the way the muscles move their skeletons. Using reference sources (textbooks, Internet, etc.), write a report comparing locomotion in insects and humans.

© 1998 J. Weston Walch, Publisher

Chemical Signals: The Scent of Life

TEACHER RESOURCE PAGE

INSTRUCTIONAL OBJECTIVES

Students will be able to

- record observations of living organisms.
- record data in a data table.
- perform calculations.
- draw conclusions based on data.
- formulate hypotheses about aspects of animal behavior.

NATIONAL SCIENCE STANDARDS ADDRESSED

Students demonstrate understanding of

- structure and function in living systems.
- regulation and behavior, such as responses to environmental stimuli.
- adaptation of organisms.
- big ideas and unifying concepts, such as cause and effect.

Students demonstrate scientific inquiry and problem-solving skills by

- identifying and controlling variables in experimental research settings.
- evaluating the accuracy, design, and outcomes of investigations.
- working in teams to collect and share information and ideas.

Students demonstrate competence with the tools and technologies of science by

- using tools such as traditional laboratory equipment to observe and measure organisms and phenomena.
- collecting and analyzing data using concepts and techniques, such as appropriate data displays.
- acquiring information in multiple sources, such as print, the Internet, and experimentation.

Students demonstrate effective scientific communication by

- representing data and results in multiple ways, such as tables and writing.
- explaining a scientific concept or procedure to other students.

MATERIALS

Each group of four students will need

- Tank of male and female Madagascan hissing cockroaches
- Two empty tanks
- Package of filter paper
- Two bottom halves of a cardboard egg carton or two peat flowerpots
- Two test tubes filled with water and plugged with moist, absorbent cotton
- Dry dog food
- Apple or potato
- Paper towels
- Petroleum jelly
- Plastic gloves
- Wax marking pencil
- Transparent plastic shoe box or similar transparent container
- Stopwatch or watch with a second hand

(continued)

Helpful Hints and Discussions

Time frame: Two single periods of instruction approximately two weeks apart
Structure: Groups of four students
Location: In class

Animals produce chemicals called pheromones to communicate with other members of their own species. In this activity, students determine how pheromones produced by *Gromphodorina portentosa,* the Madagascan hissing cockroach, affect other members of the species. These pheromones will be collected on filter paper used to line the bottom of the tank.

This activity will be done in two classroom periods approximately two weeks apart. On the first day, students will make preliminary observations and set up tanks so that pheromones may be collected. During the second lesson, students will observe the effects of these chemicals.

There are limitations to the accuracy of this activity. For example, the best way to collect the pheromones is to use methanol to extract them from feces or bedding. Because this procedure is not feasible in the average high school laboratory, an alternative is presented here. After two weeks, the pheromones present on the filter paper should produce a reaction, and students will be able to see evidences of attraction and aggression.

Gromphodorina portentosa are easy to maintain. They may be purchased from a biological supply house, such as Carolina Biological Supply Company. The Gromphs should be kept at room temperature in a screen-covered tank that has a 2-cm ring of petroleum jelly around the inside of the rim. Water should be provided in a test tube plugged with moist, absorbent cotton. Dry dog food and an occasional slice of apple or potato will supply nutrients. A small peat flowerpot or piece of egg carton will increase the living space.

Initially, some students may be squeamish about handling cockroaches. Emphasize that the cockroaches move very slowly, are completely harmless to humans, and are easy to handle. After a few tries, students will become experts at handling them, and will be very proud of themselves. Because some people may be allergic to certain insects or their waste products, it is advisable to wear gloves when touching these animals. Make sure that hands are washed thoroughly after this laboratory activity.

Adaptations for High and Low Achievers

High Achievers: These students should help lower achievers by working with them in a group. Also, encourage these students to perform additional behavioral experiments based on follow-up activity 2.

Low Achievers: Provide a glossary and/or reference material for the italicized items in this activity. Assist the groups in assigning responsibilities for timekeeping and measuring reactions. Heterogeneous groupings will enable students of higher ability to assist those having difficulty carrying out the activity.

Scoring Rubric

Full credit should be given to students who use complete sentences to correctly answer all the questions and accurately complete the three data tables provided. Extra credit should be given to students who complete any of the follow-up activities.

Internet Tie-ins

www.carolina.com (Carolina Biological Supply Co.)
www.birminghamzoo.com/so/arthro.htm (Animal Omnibus)
http://www.gene.com/ae/AE/AEC/AEF/index.html (Access Excellence)

Quiz Answer each part of the following question in one or more complete sentences.

Because insecticides have caused many environmental problems, scientists are trying to use pheromones to get rid of insect pests.

a. What are pheromones?
b. How might scientists use pheromones to catch cockroaches?
c. Why is this a good idea?
d. What problems might result from using pheromones in traps?

Name____ Date ________________________

Chemical Signals: The Scent of Life

STUDENT ACTIVITY PAGE

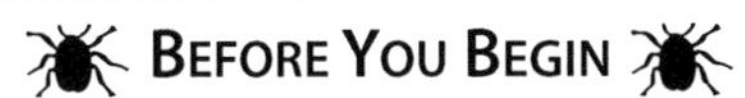

Insects communicate with members of their own *species* for many reasons. They produce signals to attract a mate, to warn others of danger, or to inform others of a potential food source or shelter. Chemicals called *pheromones* are used for signaling. In this activity, you will determine how pheromones produced by *Gromphodorina portentosa,* the Madagascan hissing cockroach, affect other members of the species.

This activity will be done in two classroom periods approximately two weeks apart. On the first day, you will make preliminary observations and set up your tanks so that pheromones may be collected. During the second lesson, you will observe their effects.

Gromphodorina portentosa are easy to maintain. They should be kept at room temperature in a screen-covered tank that has a 2-cm-thick ring of petroleum jelly around the inside of the rim. Water in a test tube plugged with moist, absorbent cotton, dry dog food, and an occasional piece of apple or potato will supply nutrients. A small peat flowerpot or piece of egg carton will increase the animals' living space.

These cockroaches move very slowly, are completely harmless to humans, and are easy to handle. Simply pick them up and place them in your hand. After a few tries, you will be able to do this like an expert and be very proud of yourself. Because some people may be allergic to certain insects, it is advisable to wear gloves when you handle these animals. Make sure that you wash your hands thoroughly after this laboratory activity.

Remember that these animals are alive and should be treated humanely. When moving the animals, be careful not to harm them and give them time to recover. Do not make any sudden noises or movements. Above all, be patient.

MATERIALS

Each group of four students will need

- Tank of male and female Madagascan hissing cockroaches
- Two empty tanks
- Package of filter paper
- Two bottom halves of a cardboard egg carton or two peat flowerpots
- Two test tubes filled with water and plugged with moist, absorbent cotton
- Dry dog food
- Slice of apple or potato
- Paper towels
- Petroleum jelly
- Plastic gloves
- Wax marking pencil
- Transparent plastic shoe box or similar transparent container
- Stopwatch or watch with a second hand

PROCEDURE

Day 1

1. Observe the cockroaches in their tank for two minutes. Take notes on your observations. Identify males and females. The male has bumpy, rounded horns on his thorax, and his antennae are

(continued)

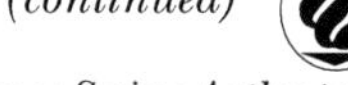

thicker and hairier than the female's (Figure 1). Do the animals appear to be communicating? Why do you think so? Look for evidences of *aggregation,* in which the animals tend to stay together. Observe their locomotion. Look for evidences of "stilt walking," in which the animal seems to be walking on stilts, with its body raised above its legs. This is an indication of aggressive behavior. How does this compare with their usual method of walking?

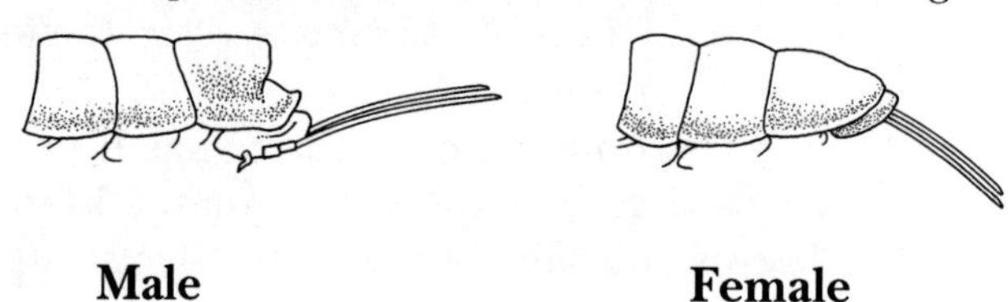

Male **Female**

Figure 1: Male and Female *Gromphodorina Portentosa*[1]

2. Label one empty tank "Male" and the other "Female." Line the bottom of each tank with six or more pieces of filter paper. Place a piece of egg carton on top of the paper. Add a test tube of water plugged with moist, absorbent cotton; a few pellets of dry dog food; and one or two slices of potato or apple. Using a paper towel, place a thick ring of petroleum jelly around the inside rim of the tank.
3. Place five of the largest males in the "Male" tank and five of the largest females in the "Female" tank.
4. Observe the males and females in their new tanks. Describe how they use their antennae to learn about their new surroundings.
5. For the next two weeks, make sure that the organisms have enough food and water. Change the apples or potatoes every few days, but *do not* clean the tanks or remove wastes.

Day 2 (2 weeks later)

1. **Read this entire section before beginning.** With other members of your group, decide how you will obtain measurements. Assign tasks to each member of your group.
2. With your wax marking pencil, draw a line down the middle of the plastic shoe box. Place one male and one female from segregated tanks at opposite sides of the box. Observe them for two minutes. With the help of other members of your group, record the amount of time they spend together and apart. Describe any interactions between the two cockroaches. Which sense organs are they using to detect each other's presence? Is there any evidence of stilt walking? Return these animals to their original tanks. Repeat your observations with another male and female pair.
3. In steps 3 through 6, you will compare the reactions of cockroaches to two pieces of filter paper, one of which contains pheromones and one of which does not. The cockroach will be exposed to these two pieces of paper for two minutes. If the animal could not detect a difference between the two pieces of paper, how much time would it spend on each piece? Why?

[1]http://www.carolina.com/tips/mar95/index.html (*Carolina Tips,* March 1995)

(continued)

© 1998 J. Weston Walch, Publisher

Name_______________________________ Date _______________

Chemical Signals: The Scent of Life *(continued)*

STUDENT ACTIVITY PAGE

Rinse the box with warm water and dry it thoroughly. Place one piece of filter paper from the female's tank on one side of the box. Place a clean piece of filter paper on the opposite side. Place a male on the line between the two pieces of paper. Observe its actions for two minutes. With the help of other members of your group, record the number of times it goes to each piece of paper. How does its reaction compare with that of the male that was in the tank with a female? Did the second animal stilt walk? Discard the filter paper and return the male to its tank.

4. Repeat step 3 with female cockroaches, using filter paper from the male's tank on one side and clean filter paper on the other.
5. Repeat step 3 with male cockroaches, using filter paper from the male's tank on one side and clean filter paper on the other.
6. Repeat step 3 with female cockroaches, using filter paper from the female's tank on one side and clean filter paper on the other.
7. When you have completed your observations, return all cockroaches to their original tank. Clean the tanks in which you kept the males and females for the last two weeks and wash your hands thoroughly.

DATA COLLECTION AND ANALYSIS

1. In the table below, record your observations of the behavior of the cockroaches on the first day. Describe any evidence of communication, aggregation, or stilt walking.

Table 1: Behavior of Cockroaches on First Day

Behavior	Tank 1: Males and Females	Tank 2: Males	Tank 3: Females
Communication			
Aggregation			
Stilt Walking			

2. In the table below, record your observations of the behavior of the male and female that were placed together after having been separated for two weeks.

Table 2: Behavior of Male and Female Cockroaches

Trial #	Number of Seconds Together	Number of Seconds Apart	Comments
Pair 1			
Pair 2			
Average			

(continued)

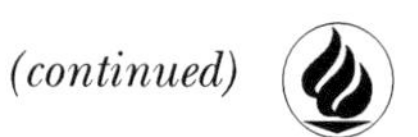

© 1998 J. Weston Walch, Publisher

Name________________________________ Date ____________________

Chemical Signals: The Scent of Life *(continued)*

STUDENT ACTIVITY PAGE

3. Complete the following table.

Table 3: Reaction to Pheromones on Filter Paper

Trial	Time (Sec) on Pheromone Paper (Experimental)	Time (Sec) on Non-Pheromone Paper (Control)	Comments
1. Male reacting to female pheromones			
2. Female reacting to male pheromones			
3. Male reacting to male pheromones			
4. Female reacting to female pheromones			

4. How did your results in Table 3 compare with your answer in question 3?
5. Draw as many conclusions about cockroach behavior as possible, based on the data in Tables 2 and 3. Write your answers in complete sentences in the space below.

CONCLUDING QUESTIONS

Answer all questions on the back of this sheet or on a separate page.

1. Why is the title of this activity a good one?
2. Why did you isolate the males and females for two weeks before testing their reactions to pheromones?
3. Why was step 2 of the procedure section necessary?
4. Why was it necessary to rinse the plastic box between trials?

FOLLOW-UP ACTIVITIES

1. By the end of two weeks, there are usually stains on the filter paper that has been in the tank with cockroaches. Design an experiment to discover whether the cockroaches are reacting to visual or olfactory stimuli. Discuss your experiment with your teacher and then carry it out.
2. Pheromones have recently been used as biological controls of various insects. Use books, journals, or the Internet to learn more about this. Report your findings to the class.

Tribolium Confusum: How Can You Confuse This Flour Beetle?

TEACHER RESOURCE PAGE

INSTRUCTIONAL OBJECTIVES

Students will be able to

- record observations of a living organism.
- formulate hypotheses.
- record data in a data table.
- draw conclusions based on data.
- draw and label *Tribolium confusum* life cycle stages.

NATIONAL SCIENCE STANDARDS ADDRESSED

Students demonstrate understanding of

- reproduction and heredity.
- regulation and behavior.
- populations and ecosystems.
- adaptations of organisms.
- big ideas and unifying concepts, such as cause and effect.

Students demonstrate scientific inquiry and problem-solving skills by

- identifying or controlling variables in experimental research settings.
- proposing, recognizing, analyzing, and critiquing alternative explanations.
- evaluating the accuracy, design, and outcomes of investigations.
- working in teams to collect and share information and ideas.

Students demonstrate competence with the tools and technologies of science by

- using technology and tools (such as traditional laboratory equipment) to observe organisms and phenomena.
- recording and storing data using a variety of formats.
- acquiring information from multiple sources, such as print, the Internet, and experimentation.

Students demonstrate effective scientific communication by

- representing data and results in multiple ways, such as numbers, tables, graphs, drawings, and writing.
- arguing from evidence, such as data produced through their own experimentation.

MATERIALS

Each group of four students will need

- Established culture of *Tribolium confusum* containing all stages of their life cycle
- Two empty plastic containers at least one gallon in size, covered with mesh
- Scoopula or spoon
- Glass-marking pencil
- Covered petri dish
- Stereomicroscope
- Flour
- Brewer's yeast
- Potato
- Paper towels
- Wheat germ
- Triple-beam balance
- Weighing paper
- Small paintbrush or *Drosophila* sorting brush

(continued)

HELPFUL HINTS AND DISCUSSION

Time frame: Six single periods of instruction approximately five days apart
Structure: Groups of four students
Location: In class

In this activity, students examine conditions that may lead to cannibalistic behavior in the flour beetle, *Tribolium confusum.* This insect pest, which may be found in packaged flour and other grains, is small (approximately 5 mm in length) and easy to maintain.

Tribolium may be purchased from biological supply houses such as Carolina Biological Supply Company. They may be kept at room temperature in the laboratory in a plastic container at least one gallon in size. A fine-mesh cover will provide ventilation and prevent the organisms from escaping. A 10:1 ratio of wheat flour to brewer's yeast will provide nutrition. Moisture may be obtained from a few chunks of raw potato. Shredded paper towels on top of the flour will serve as an egg-laying area.

Although these animals are not harmful to humans, they certainly are pests if they get into food supplies. Therefore, warn your students to be very careful not to let any escape. Be sure that all organisms are returned to their containers at the end of the activity and that the covers are securely fastened.

ADAPTATIONS FOR HIGH AND LOW ACHIEVERS

High Achievers: These students should help lower achievers by working with them in a group. Also, encourage these students to perform additional experiments based on concluding question 3 or any of the follow-up activities.

Low Achievers: Provide a glossary and/or reference material for the italicized terms in this activity. Heterogeneous groupings will enable students of higher ability to assist those having difficulty carrying out the activity. Encourage these students to do follow-up activity 4.

SCORING RUBRIC

Full credit should be given to students who use complete sentences to correctly answer all the questions and accurately complete the three data tables provided. Extra credit should be given to students who complete any of the follow-up activities.

www.carolina.com (Carolina Biological Supply Co.)
www.colostate.edu/depts/entomology/ent.htm

Answer the following questions in complete sentences.

1. In which stage of its life cycle will the flour beetle cause the *least* damage? Why?
2. The adult flour beetle produces 400–500 eggs during its life. Explain why there is not a population explosion of these animals.

Name____ Date ______________________________

Tribolium Confusum: How Can You Confuse This Flour Beetle?

STUDENT ACTIVITY PAGE

The flour beetle, *Tribolium confusum,* is an insect pest that may be found in packages of flour and other grains. Because they are small (approximately 5 mm in length) and hardy, they may be particularly bothersome. They are *cannibalistic,* which is unusual behavior for insects. The larvae may feed on *Tribolium* eggs, and the adults eat their own *larvae* and *pupae.* In this activity, you will determine conditions that may lead to such cannibalistic behavior.

Tribolium should be kept at room temperature in a plastic container that will hold at least one gallon. A fine-mesh cover will provide ventilation and prevent the organisms from escaping. A 10:1 ratio of wheat flour to brewer's yeast will provide nutrition, and a few chunks of raw potato will provide moisture. Shredded paper towels on top of the flour will serve as an egg-laying site.

Although these animals are not harmful to humans, they certainly are pests if they get into food supplies. Therefore, be very careful when working with them. Make sure that all organisms are returned to their containers at the end of the activity and that the covers are securely fastened.

- Established culture of *Tribolium confusum* containing all stages of their life cycle
- Two empty plastic containers at least one gallon in size, covered with mesh
- Scoopula or spoon
- Glass-marking pencil
- Covered petri dish
- Stereomicroscope
- Flour
- Brewer's yeast
- Potato
- Paper towels
- Wheat germ
- Triple-beam balance
- Weighing paper
- Small paintbrush or *Drosophila* sorting brush

PROCEDURE

1. Observe the organisms in their original container for two minutes. Look for adults, eggs, larvae, and pupae. How many stages of their life cycle can you see? Draw the stages and label as many structures as possible in the Data Collection and Analysis section.
2. With your scoopula or spoon, remove some of the organisms and place them in a petri dish. Cover the dish. Observe them under the stereomicroscope. What additional organisms and/or structures do you see? Draw the organisms and label as many structures as possible.
3. Prepare two containers in the following manner:

   Container 1 (labeled *Control)*: Place 60 grams of flour, 6 grams of yeast, and 4 slices of raw potato into the container. Add five strips of paper towels.

   Container 2 (labeled *Experimental)*: In addition to the ingredients in container 1, add 20 grams of wheat germ.

   Hypothesize in which container the beetles will become cannibalistic.

(continued)

Name______________________________ Date ______________

Tribolium Confusum: How Can You Confuse This Flour Beetle? *(continued)*

STUDENT ACTIVITY PAGE

4. Look for pupae. These are light brown, almost white *sessile* organisms. Observe them under the stereomicroscope. With the help of Figure 1, identify males and females. Use your sorting brush to move 10 males to one side and 10 females to the other.

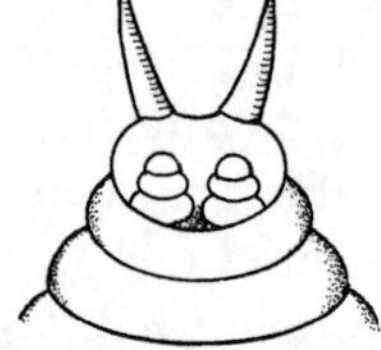

Female **Male**

Figure 1: Terminal Segments of Flour Beetle Pupae[1]

5. Place five male and five female pupae into each container. Cover the containers. These containers may be left in the laboratory for the duration of this activity (approximately one month). Once a week, add two new potato slices to each container. Do not remove the old slices.
6. On day 5, record the number of pupae and adults in each container. Also look for any signs of cannibalism, such as missing parts. On day 10, in addition to recording the number of pupae and adults, look for evidence of eggs and larvae on the paper and potato. For more accurate observations, place the paper and potato slices in a covered petri dish and observe them under the stereomicroscope. Count the number of larvae on two potato slices in each container and determine the average. Look for evidence of cannibalism.
7. On days 15, 20, and 25, repeat step 6.
8. When you finish this activity, follow your teacher's instructions for cleaning up.

DATA COLLECTION AND ANALYSIS

1. Complete the following chart by drawing the organisms.

Table 1: Life Cycle of *Tribolium Confusum*

Stage	As Seen with Naked Eye	As Seen Under Stereo Microscope (3)
Egg		
Larva		
Pupa		
Adult		

[1]*Carolina Arthropods Manual.* (Burlington, North Carolina: Carolina Biological Supply Co., 1982)

(continued)

Name________________________________ Date ________________

Tribolium Confusum: How Can You Confuse This Flour Beetle? *(continued)*

STUDENT ACTIVITY PAGE

2. Complete the following tables, based on your observations.

Table 2: Changes in the Flour Beetle Population (Control Group)

Day	Eggs	Larvae	Pupae	Adults
1	0	0	10	0
5				
10				
15				
20				
25				

Table 3: Changes in the Flour Beetle Population (Experimental Group)

Day	Eggs	Larvae	Pupae	Adults
1	0	0	10	0
5				
10				
15				
20				
25				

3. In which container did you observe evidence of cannibalism? Why do you think the flour beetles in this container became cannibalistic?
4. Write a paragraph explaining whether your results agree with your hypothesis. Use information from the data tables and your observations to support your answer.

CONCLUDING QUESTIONS

Answer all questions on the back of this sheet or on a separate

1. Based on what you learned in this activity, why do you think that *Tribolium confusum* is said to be confused?
2. How can you prevent foods such as flour or cereals from becoming contaminated with these insects?

(continued)

© 1998 J. Weston Walch, Publisher

Tribolium Confusum: How Can You Confuse This Flour Beetle? *(continued)*

3. Any differences that you may have observed between your control and experimental groups may have resulted from either the addition of wheat germ or the addition of an extra 20 grams of material. How could the experiment be improved to determine which of these factors caused the difference?

FOLLOW-UP ACTIVITIES

1. Use the data from Tables 2 and 3 to produce a line graph titled Number of *Tribolium Confusum* Larvae. Label the *x*-axis (horizontal axis) "Day." Label the *y*-axis (vertical axis) "Number of Larvae". Use black ink for the control line and red ink for the experimental line.
2. Some people store flour in the refrigerator. Design an experiment to determine how refrigeration would affect *Tribolium confusum,* which may be found in flour. After discussing your ideas with your teacher, perform the experiment.
3. Using books, journals, or the Internet, learn more about other beetles, such as *Tenebrio* or ladybird beetles. What characteristics do they share? Describe the role that some of these beetles play in the environment. Present your findings to the class.
4. Using books, journals, the Internet, or information on labels, prepare a report on the nutritional value of flour and wheat germ. Present the report to your class.

© 1998 J. Weston Walch, Publisher

Controlling Insects: Natural Repellents

TEACHER RESOURCE PAGE

INSTRUCTIONAL OBJECTIVES

Students will be able to

- record observations of a living organism.
- record data in a data table.
- draw conclusions based on data.
- evaluate reliability of data.

NATIONAL SCIENCE STANDARDS ADDRESSED

Students demonstrate understanding of

- regulation and behavior, such as response to environmental stimuli.
- big ideas and unifying concepts, such as cause and effect.
- the designed world, such as the development of agricultural techniques.
- personal and environmental safety.

Students demonstrate scientific inquiry and problem-solving skills by

- identifying or controlling variables in experimental settings.
- using evidence from reliable sources to develop descriptions, explanations, and models.
- distinguishing between fact and opinion.
- evaluating the accuracy, design, and outcomes of investigations.
- working individually and in teams to collect and share information and ideas.

Students demonstrate competence with the tools and technologies of science by

- using tools to observe and measure organisms and phenomena.
- recording and storing data.
- acquiring information from multiple sources, such as print, the Internet, and experimentation.

Students demonstrate effective scientific communication by

- arguing from evidence, such as data produced through their own experimentation.
- critiquing published materials.

MATERIALS

Each group of four students will need

- Tank containing *Tenebrio* larvae, bran flakes, and a test tube of water covered with moist, absorbent cotton
- Small empty tank lined with graph paper or a grid
- Several pieces of filter paper
- Plastic gloves
- Stopwatch or watch with a second hand
- Metric ruler
- Substances that might be used as repellents, such as slices of garlic, cucumber skins, hand lotion, perfume, coffee grounds, or marigolds

(continued)

Helpful Hints and Discussion

Time frame: Single period of instruction
Structure: Groups of four students
Location: In class

In this activity, students test the effectiveness of natural insect repellents on the larvae of the flour beetle, *Tenebrio molitar.* These inexpensive organisms are readily available from biological supply houses and are easy to maintain in the classroom. They may be kept in a screened tank that contains a layer of wheat bran, a slice of potato or apple, and a test tube of water covered with moist, absorbent cotton.

Because *Tenebrio* move rather slowly, you might want to use a small tank, such as a children's shoe box, for your experimental container. To simplify measurements, line the tank with a grid that contains dark lines 1 cm apart.

When introducing this activity, emphasize that many commercial insecticides and repellents are harmful to the ecosystem. Charge your students with the responsibility of finding a repellent that will not harm any organisms. In addition to the substances mentioned above, they might want to try some favorite family remedies.

Remind your students that *Tenebrio* are living organisms and should be handled carefully and treated humanely. To avoid allergic reactions, plastic gloves should be worn when handling *Tenebrio.* Hands should be washed thoroughly after the experiment.

Adaptations for High and Low Achievers

High Achievers: These students should help lower achievers by working with them in a group. Also, encourage these students to perform additional experiments based on follow-up activity 1 or 2.

Low Achievers: Provide a glossary and/or reference material for the italicized terms in this activity. Heterogeneous groupings will enable students of higher ability to assist those having difficulty carrying out the activity. Encourage them to do follow-up activity 3.

Scoring Rubric:

Full credit should be given to students who use complete sentences to correctly answer all the questions and accurately complete the data table provided. Extra credit should be given to students who complete any of the follow-up activities.

Internet Tie-ins

www.birminghamzoo.com/ao/arthrop.htm (Animal Omnibus)
www.colostate.edu/Depts/Entomology/ent.htm

Quiz

Use at least one complete sentence to answer each of the following questions.

1. How does an *insecticide* differ from an *insect repellent?*
2. Why are many people concerned about the use of insecticides?

Name ____________________ Date ____________________

Controlling Insects: Natural Repellents

STUDENT ACTIVITY PAGE

Before You Begin

There are many insect pests that bite us, destroy our crops, or otherwise interfere with our lives. To solve the problems caused by these animals, we have developed various *insecticides* and *repellents.* Unfortunately, many of these are also harmful to humans and other organisms. There are, however, repellents that keep insects away but do not harm other organisms. You have probably heard stories about certain lotions or herbs or chemicals that protect against various insects. In this activity, you will test some substances that might be used as natural repellents and decide whether they are effective against the mealworm, *Tenebrio molitar.*

When handling the animals, be gentle. They move slowly and do not bite, so it will be easy to lift them with your hands. Practice a few times before taking them out of the tank. Because some people may be allergic to certain insects, it is advisable to wear gloves when you touch them. Make sure that you wash your hands thoroughly after this laboratory activity.

Materials

Each group of four students will need

- Tank containing *Tenebrio* larvae, bran flakes, and a test tube of water covered with moist, absorbent cotton
- Small empty tank lined with graph paper or a grid
- Several pieces of filter paper
- Plastic gloves
- Stopwatch or watch with a second hand
- Metric ruler
- Substances that might be used as repellents, such as slices of garlic, cucumber skins, hand lotion, perfume, coffee grounds, or marigolds

Procedure

You will be working in groups of four. This will enable you to share your observations. You will learn more if you encourage input from all the members of the group. Work as a team to answer the questions and to record your data and answers in the Data Collection and Analysis section.

1. Place a piece of moist filter paper in one corner of the empty tank. Add five mealworms to the opposite side of the tank. Observe them for two minutes. Count the number of animals that reached the filter paper. What did they do when they got there? If all the animals did not reach the paper, how far away was each one?
2. Remove the animals and the filter paper from the tank. Place a new piece of moist filter paper with two slices of cut garlic on it in the same corner as the first. Add the mealworms. Record your observations as in step 1.
3. Repeat step 2 several more times, using a different substance each time. For example, you may use cucumber skins, hand lotion, or coffee grounds. Record your observations.
4. When you have completed your observations, clean up your station and return the animals to their original tank.

(continued)

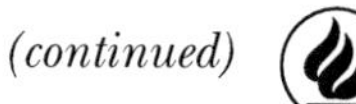

© 1998 J. Weston Walch, Publisher

Name ______________________________ Date ______________

Controlling Insects: Natural Repellents *(continued)*

STUDENT ACTIVITY PAGE

DATA COLLECTION AND ANALYSIS

1. Record your observations in the table below.

Substance on Moist Paper	# of Tenebrio Reaching Paper	Maximum Distance Away from Paper	Notes on Behavior
Nothing			
Garlic			

2. Why was it necessary that you not put anything on the moist filter paper in step 1?
3. Why was it necessary to change the filter paper and remove the animals from the tank after each substance was tested?

CONCLUDING QUESTIONS

Answer all questions on the back of this sheet or on a separate page.

1. Compare your group's results with those of other groups. What similarities and differences do you notice? How can you explain any differences?
2. Why was it important to use more than one *Tenebrio* in each test?
3. Explain whether any of the substances you tested might be used as a *Tenebrio* repellent. Why or why not?

FOLLOW-UP ACTIVITIES

1. Repeat this activity, changing the order in which you test each substance. Why might the results be different?
2. Repeat this activity, using the adult *Tenebrio* (flour beetle) instead of the larva.
3. Using the library or the Internet to find material, write a report on insecticides or insect repellents.
4. You find an advertisement on the Internet that states that a certain brand of perfume will keep insects from biting you. What will you need to know about this perfume before you buy it?

Share Your Bright Ideas

We want to hear from you!

Your name________________________________Date______________

School name____

School address____

City ____________________State_____Zip__________Phone number (______)__________

Grade level(s) taught________Subject area(s) taught______________________

Where did you purchase this publication?______________________________

In what month do you purchase a majority of your supplements?____________________

What moneys were used to purchase this product?

___School supplemental budget ___Federal/state funding ___Personal

Please "grade" this Walch publication in the following areas:

Quality of service you received when purchasing	A	B	C	D
Ease of use	A	B	C	D
Quality of content	A	B	C	D
Page layout	A	B	C	D
Organization of material	A	B	C	D
Suitability for grade level	A	B	C	D
Instructional value	A	B	C	D

COMMENTS:____

____

What specific supplemental materials would help you meet your current—or future—instructional needs?

____

Have you used other Walch publications? If so, which ones?______________________

May we use your comments in upcoming communications? ___Yes ___No

Please **FAX** this completed form to **888-991-5755**, or mail it to

Customer Service, Walch Publishing, P. O. Box 658, Portland, ME 04104-0658

We will send you a **FREE GIFT** in appreciation of your feedback. **THANK YOU!**